John Brockman is the author/editor of nineteen books. He is founder of Brockman Inc., a literary and software agency, founder of The Reality Club, president of Edge Foundation Inc. and editor and publishe leading scientists and thinker

The Greatest Inventions of the Past 2,000 Years

EDITED BY JOHN BROCKMAN

PHŒNIX

A PHOENIX PAPERBACK

First published in Great Britain in 2000
by Weidenfeld & Nicolson
This edition is published in association with Simon & Schuster,
Rockefeller Center, 1230 Avenue of the Americas,
New York, NY 10020

This paperback edition published in 2001
by Phoenix,
an imprint of Orion Books Ltd,
Orion House, 5 Upper St Martin's Lane,
London WC2H 9EA

'Afterword' by Jared Diamond is an expanded
version of an article that appeared in
the *New York Times* magazine.

A CIP catalogue record for this book
is available from the British Library.

ISBN 0 75381 128 6

Printed and bound in Great Britain by
The Guernsey Press Co. Ltd, Guernsey, C.I.

ACKNOWLEDGMENTS

Anthony Cheetham, chairman of the Orion Publishing Group in London, was the first to suggest that the *Edge* inventions poll would make a valuable book, and I thank him for his wholehearted goodwill and encouragement. On this side of the Atlantic, Bob Bender at Simon & Schuster was enthusiastic and unfailing in his help. I am also indebted to Sara Lippincott for her thoughtful editing. I owe most, of course, to the *Edge* poll participants themselves, and am most grateful for their insightful contributions.

To my brother,
the physicist Philip Brockman

Contents

PREFACE

As a follow-up to my book *The Third Culture* (1995), in January 1997 I launched *Edge* (www.edge.org), a website displaying the activities of an invitation-only "third-culture mail list."

The third culture consists of those scientists and other thinkers in the empirical world who, through their work and expository writing, are taking the place of the traditional intellectual in redefining who and what we are. *Edge* gives these thinkers an opportunity to communicate their ideas not just to the public but to one another, with the understanding that those ideas will be challenged. *Edge*'s motto is: "To arrive at the edge of the world's knowledge, seek out the most complex and sophisticated minds, put them in a room together, and have them ask each other the questions they are asking themselves."

The ideas presented on *Edge* are often highly speculative; they represent the frontier of knowledge in the areas of evolutionary biology, genetics, computer science, neurophysiology, psychology, physics. Some of the fundamental questions that have been posed are: Where did the universe come from? Where did life come from? Where did the mind come from?

Arising from the third culture is a new natural philosophy

founded on reconceptions of complexity and of evolution. A new set of metaphors has emerged to describe ourselves, our minds, the universe, and all of the things we know in it; and it is the intellectuals concerned with these new ideas and images—scientists writing books about their own work—who drive our times. Featured to date on *Edge* have been such third-culture scientists as the evolutionary biologist Richard Dawkins, the philosopher Daniel C. Dennett, the physiologist Jared Diamond, the computer scientist Marvin Minsky, and the psychologist Steven Pinker.

In November 1998, I put out an e-mail query to the mail list, asking for their responses to the following questions: "What Is the Most Important Invention in the Past Two Thousand Years?" and "Why?"—for an article that was subsequently published in *Edge* 48, on January 4, 1999. The robust and sometimes amusing electronic discussion that ensued attracted an unusual degree of attention in other media (ABC News, the *Wall Street Journal, Newsweek,* the *Daily Telegraph*), along with the interest of my publishers. It was decided that the observances accompanying the arrival of the new millennium would not be complete without some survey, by people who think a lot about technology and where it is taking all of us, of the major technological advances in what's known in the West as the Common Era. Responses came in from a great many scientists, philosophers, and entrepreneurs, and I have selected a hundred or so of these responses for inclusion in this book—with an unavoidable arbitrariness, for which I won't apologize.

In such a compendium, a conventional table of contents is necessarily something of a lengthy and dizzying nuisance; readers may plunge in and march straight through, or browse at will. They may consult the index of contributors, if they wish, to locate entries by their favorite thinkers.

Of course, I have ideas of my own on The Most Important Invention, and here they are.

Early in my career, I had the good fortune to work as a producer and impresario with a number of artists who were focusing on the new ideas and metaphors coming out of the work of, among others, Norbert Wiener in cybernetics, Claude Shannon in information theory, and Marshall McLuhan in communications.

Entering composer LaMonte Young's Theatre of Eternal Music, an environment in which one chord was played for hours (or days) on end, I was immersed in an aural space in which my sensory mechanisms became the tuning mechanism for his composition. Screening Andy Warhol's eight-hour film *Empire* (and sitting through it in its entirety), while admittedly not something I would want to do on a regular basis, provided me with a unique and lasting impression of the nature of time and boredom. At the first intermedia discothèque, Murray the K's World, which I conceptualized and helped produce in an old airplane hangar at Roosevelt Field on Long Island in 1967, the environment created by Gerd Stern's USCO Group—hundreds of pulsating audio and video inputs simultaneously massaging attendees from all sides—helped me to understand the cybernetic idea of feedback as a nonlinear relationship between output and input. The words "information" and "communication" took on new and radical meanings as indicators of the ways in which minds affected one another.

A couple of years earlier, Gerd Stern had toured college campuses with the Canadian communications guru Marshall McLuhan. It was an odd juxtaposition: Stern and the USCO Group were perhaps best known for the "Psychedelic Tabernacle" they had set up in a refurbished church in Garnerville, New York; in

their tie-dyed clothes they would stage an intermedia perfor-
mance, which was then followed by a talk by the very conserva-
tively dressed professor from Toronto. McLuhan had famously
remarked, in his 1964 book *Understanding Media: The Extensions
of Man*, that in inventing the computer we had externalized our
central nervous systems—that is, our minds. His collaborator Ed-
mund Carpenter, an anthropologist who studied the aural environments of North American Eskimos, invited me to lunch at
Fordham in 1967 to talk about Carpenter's ideas on acoustic space.
There I met McLuhan for the first time, and I remember his observation that we had gone beyond Freud's invention of the unconscious: for the first time we had rendered the conscious visible.

The composer John Cage used to talk about these ideas at a series of dinners in the mid-sixties at the home of Dick Higgins, cofounder of the artists' group Fluxus. Cage held that there was
"only one mind, the one we all share." Synthesizing McLuhan and
Zen, he believed that mind had become socialized, and that as a
consequence "we can't change our minds without changing the
world." Cage saw mind as the new environment; he called it "the
collective consciousness," and he maintained that we could tap
into this new environment by creating "a global utilities network."

In this new epistemology, Ockham's razor meets Gödel's theorem, and the fabric of our habitual thinking is torn apart. Subject
and object fuse. Reality passes into description—and thus becomes
invention. Such ideas may appear destructive; in fact they are liberating, allowing us to lay waste to the generalizations of previous
epochs, which we de-create by subsuming the history of our
words. As Wallace Stevens put it, "The words of the world are the
life of the world. It is the speech of truth in its true solitude: a nature that is created in what it says."

"Reality" is not about the interaction of parts of the universe;
it's about a universe that interacts infinitely with itself, where im-

portance lies in the patterns that connect the parts. Any analysis of this new state of affairs with available language will lead most people into a spiraling ladder of recursive mirrors. There is no "answer" to this conundrum; the description of the plane of language is the plane that holds our descriptions. Language becomes a commission, a dance, a play, a song.

I propose, as the most important invention of the past two thousand years, our Distributed Networked Intelligence (DNI). DNI is the collective, externalized mind—the mind we all share, the infinite oscillation of our collective consciousness interacting with itself, becoming aware of itself, adding a fuller, richer dimension to what it means to be human.

BETHLEHEM, CONNECTICUT
JUNE 1999

THE
GREATEST INVENTIONS
OF THE
PAST 2,000 YEARS

PART I

How We Live

BRIAN C. GOODWIN

The Printing Press

The most important invention of the past two thousand years is the printing press. When William Caxton published *The Canterbury Tales* on his newly invented printing machine in the fifteenth century, he dramatically accelerated the separation of human culture from nature, eclipsing the direct experience of natural processes that continues in the oral tradition and replacing it with words on a page.

This innovation cut in two directions. We could begin the systematic accumulation of knowledge in order to control nature, thus liberating ourselves from drudgery and freeing our imaginations. At the same time, nature came to be seen as little more than a set of mechanisms that we could manipulate for our own purposes; as a consequence, the "rape of nature" began in earnest.

We are now reaping twin harvests: the vastly expanded potential for written communication through the Internet (as in this exchange of views at the *Edge* website) and a vastly degraded planet that, as things are going, won't support us much longer. We are all aware of global warming, chemical pollution, the degradation of essential life support systems, the extinction of species, and human overpopulation of the planet as palpable signs of crisis.

Can we use the one harvest to save us from the other? We can now connect with our fellow human beings as never before. But what about nature? Can we use our knowledge to restore the bonds of mutual dependence between ourselves and the other members of our planetary society, so that we cease to be exploiters and become responsible participants in the creative drama in which we are all engaged, using our imaginations to free ourselves from a cultural distortion that arose from the replacement of nature by symbols on a page?

BRIAN C. GOODWIN is a professor of biology at Schumacher College, Dartington, in Devon, England, where he coordinates a master's program in holistic science. His books include *Analytical Physiology of Cells and Developing Organisms; How the Leopard Changed Its Spots: The Evolution of Complexity;* and (with Gerry Webster) *Form and Transformation: Generative and Relational Principles in Biology.*

RODNEY BROOKS

The Electric Motor

My answer is the electric motor, in all its guises in which electricity produces mechanical motion.

The industrial revolution was restricted to places of work and of shared production, until the relatively small and clean electric motor enabled the adoption of its bounty into the home: refrigeration, automated cleaning, cooling, better heating, entertainment, mass data storage, home medical care, and more comfortable personal transportation.

True, many of these boons were already present in the home thanks to simpler technologies—gravity-driven plumbing, for example, or convective air flow—but it was the electric motor that made them pervasive. The transformation of our Western lifestyle has been profound and has completely changed our expectations of how our bodies should fit with our surroundings.

A question: What will it take for the computer revolution to enter our lives in the way that the electric motor has enabled the industrial revolution to do?

RODNEY BROOKS, a computer scientist, is the director of MIT's Artificial Intelligence Laboratory; the author of *Model-Based Computer Vision, Programming in Common LISP*, and *Cambrian Intelligence: The Early History of the New AI;* and co-editor (with Luc Steels) of *The Artificial Life Route to Artificial Intelligence.*

TOM STANDAGE

Telecommunications Technology

I t all depends on how you define "important," of course. But to my mind the most important invention is telecommunications technology: the telegraph, the telephone, and now the Internet. Until about a hundred and fifty years ago, it was impossible to communicate with someone in real time unless that person was in the same room. The only options were to send a message (or go yourself) by horse or ship.

The early optical telegraphs of the 1790s made long-distance communication possible at hitherto impossible speeds, at least for the governments that built them; but they were not available for general use. Then, in the 1840s, the introduction of the electric telegraph enabled people to send messages over great distances very quickly. This was a step change, though its social consequences took a while to percolate. At first, telegraph operators became the pioneers of a new frontier—they could get together in what we would today call chat rooms, play games over the wires, and so on. There were several telegraphic romances and weddings. The general public, of course, was still excluded and had no direct access to the real-time nature of the technology until the invention of the telephone in the 1870s.

Today—in the developed world, at least—we think nothing of talking with people on the other side of the globe. In the course of a normal working day, many people spend more time dealing with people remotely than they do face-to-face. The ubiquity of telecommunications technology has become deeply embedded in our culture. We watch TV and use telephones, fax machines, and, increasingly, the Internet, almost unthinkingly. If the mark of an advanced technology is that it is indistinguishable from magic, then the mark of an important technology is that it becomes invisible—that we fail to notice when we are using it. The significance of telecommunications technology is thus very easy to overlook and underestimate.

Tom Standage, the science correspondent of the *Economist* and the former deputy editor of the *Daily Telegraph*'s technology supplement, "Connected," is the author of *The Victorian Internet: The Remarkable Story of the Telegraph and the Nineteenth Century's On-line Pioneers.*

COLIN TUDGE

The Plow

Why do we always think of bears, wolves, rhinos, and tigers as inhabiting the wildest and woolliest places—mountain, forest, desert, and swamp? Why should we not see them, in our mind's eye, on the pleasant slopes of Berkshire or in the Napa Valley? Because, of course, those easy places have long since succumbed to agriculture, and wild species have mostly been marginalized. How could this have happened on such a scale—worldwide—and so quickly? Primarily because of just one invention: the plow. Human beings have probably practiced farming of a kind for at least thirty thousand years, but it was not until they began to break the soil systematically that they truly began to dominate the landscape and produce crops on what for many a century has been an industrial scale. The so-called neolithic revolution of ten thousand years ago probably represents the start of this—not the beginnings of agriculture, but the start of agriculture on a large scale. The plow has enabled human beings to break the ecological law which says that big, fierce animals are rare, for we are the most formidable of all, and yet there are now six billion of us. The plow has transformed the world's landscape and pushed our fellow species to the sidelines. The first plows easily predate this book's prescribed time frame, the last two thousand years, but the plow has come on a great deal in that time, and a device this significant just has to be included.

Farming, defined broadly, is the management of environment in ways that increase the human food supply. There are many ways to do this without breaking the soil *in extenso*, ways that surely were practiced long before the neolithic revolution. Cultivation of individual plants is *horticulture* (from the Latin *hortus*, for "garden") and includes crop protection (driving away the monkeys as well as the insects), weeding, and pruning, along with propagation. Management of animals is *pastoralism,* and there are also many forms of herding on the way to domestication; Australian aborigines control kangaroos by setting judicious bush fires, for instance. Such proto-farming makes a huge difference to human well-being, but its output is always limited, and it leaves the general landscape more or less intact.

But arable farmers take the landscape by the scruff and bend it to their will. The native flora is stripped back to the bare soil, a mere substrate. The original landscape is irrelevant. Many scholars have suggested that the banishment of Adam and Eve from Eden represents the beginnings of large-scale agriculture: people had overhunted the local deer and gazelles, and the rising seas at the end of the last Ice Age had drowned the coastal plains, forcing people who had previously managed the local flora for a hobby to cultivate full-time. I believe that Cain's murder of Abel is emblematic of the start of arable farming; the "tiller of the ground" kills the gentle shepherd. The authors of Genesis clearly acknowledge the significance of this. For them the breaking of the ground, the obliteration of God's own landscape, was blasphemy. The furrowed soil is truly the mark of Cain. The ancient conflict of arable farmer and pastoralist persisted deep into the twentieth century—reflected in the song from *Oklahoma:* "Oh, the farmer and the cowman should be frie-e-e-ends!"

The first plows were forked sticks drawn by human power—surely horrendous labor, like most of arable agriculture until recent times. For all the effort, though, these plows merely scratched the ground. Advance since then has taken three forms: the use of stronger and more versatile materials—first wood capped by iron, then iron alone, then steel; the introduction of more and more subtle designs—first the share was added to break the soil, and then the moldboard, which turns it over; and finally, the increase in traction power, on which all other advances depended. Human gave way to ox power long before the time of Christ, but the heavy horses that came later were faster and even stronger (although they ate more and so were less thrifty). The Chinese invented the collar that enabled horses to pull without strangling themselves. But it was the tractor that truly pushed the world into its modern phase. Tractors partake of Isambard Kingdom Brunel's philosophy of engineering: if a given force won't do the job, then multiply by ten and try again. The heaviest clay soils could not be plowed by horses, but tractors can do anything. Thus the English county of Lincolnshire was committed to sheep grazing for two thousand years; but now, with modern tractors, it is border-to-border cereal and potato country. America's prairies show the same phenomenon on a vastly greater scale.

Forget the computer and the telegraph, forget writing, forget even the wheel. It's the plow that has changed the world.

COLIN TUDGE is a research fellow of the Centre for Philosophy at the London School of Economics. His recent books include *Last Animals at the Zoo*, *The Engineer in the Garden*, *The Time Before History*, and *Neanderthals, Bandits, and Farmers*.

ARNOLD TREHUB

Otto von Guericke's Static Electricity Machine

The most important invention in the past two thousand years must be a seminal invention with the broadest and most significant consequences. In my opinion, it is the invention by Otto von Guericke in 1660 of a machine that produced static electricity. Although electrical and magnetic phenomena were noted and commented on many centuries earlier, von Guericke's invention was *the first machine to produce electricity*. This device was the primitive tool that unlocked our understanding and application of electricity. Improvements and elaborations of the von Guericke machine followed fairly rapidly over the next several centuries, contributing to a growing understanding of electricity and its practical utilization. Modern power generation, communication, computation, transportation, and almost all of our most important analytic devices stand on the foundation of von Guericke's machine. At the level of everyday living, it is hard to imagine any of our manufacturing facilities or our households without electricity. A long line of basic intellectual formulations, from electromagnetism to the bioelectric properties of brain mechanisms, owe a debt to this invention. When we discover how the human brain creates the covert models of its own inventions, the structure and dynamics of the brain's own electrical activity will undoubtedly be an essential aspect of the explanation.

ARNOLD TREHUB is adjunct professor of psychology at the University of Massachusetts at Amherst and the author of *The Cognitive Brain*.

ALUN ANDERSON

The Caravel

The Portuguese caravel of the early fifteenth century looks at first sight more like a misbegotten hybrid of the sailing ships of two continents than the medieval equivalent of the Apollo space capsule. But Apollo is its closest relative in more ways than one. Apollo changed our world by giving us a moon's-eye view of it. The caravel changed the medieval world far more, as it crept around the shores of Africa, encountered unknown lands, and made way for the Age of Discovery, led by Vasco da Gama and Christopher Columbus. Like the Apollo program, the building and voyages of the caravels were hideously expensive; they eventually drained away to nothing the enormous wealth of their principal backer, Prince Henry of Portugal (Henry the Navigator, 1394–1460). And like Apollo, the early caravels were built as compact as they could be for their immense task of exploration, then every bit as difficult as reaching the moon.

There are no surviving caravels, nor even any detailed plans of them. The Portuguese carpenters who built the caravels were prohibited from explaining how they were constructed or from selling ships to foreigners. But we can get a pretty good idea of how they looked from surviving sketches. The body of the vessel is reminiscent of the small galleons you see in pirate movies for children, minus the rows of guns belowdecks. But the sails are not the square rig you might expect. Instead, they are triangular and hung from a sloping beam, like the rigging of the Arabian dhows that still sail the Indian Ocean, except that the caravel had two, three, and later four masts.

The caravel was indeed a hybrid of European and Arabian influences. With its triangular sails drawn in tight like those of a modern racing yacht, it could sail at an angle close to the wind instead of simply being blown along like the early square-rigger. At best, square-rigged sails of the period could fill only when they were more than 65 degrees off the wind, and square-riggers were condemned either to seek out following winds or forever tack impotently back and forth.

The caravel design was perfected in the early to mid-fifteenth century during a series of voyages financed by Prince Henry. Portugal was then at the far edge of the known world, thousands of miles overland from the wealth of Asia and the spice trade, and Henry's ambition was in part to find a southern route to Asia by going beyond the known limits of the North African coast. (In part, too, Henry may have hoped to find the fabled Christian kingdom of Prester John.) To attempt this mission, not only did the boats have to be improved, but so too did navigation, cartography, astronomy, and the knowledge of winds and tides—and so they were, in a vastly expensive program centered on the school of navigation that Prince Henry established at Sagres.

The caravels were only some sixty to one hundred feet long. They did not have space for substantial cargo, as trade was not their immediate goal; they were vessels of exploration—the maritime equivalent of a space probe. They were built with tightly sealed decks and small hatches and coated with pitch to make them as watertight as possible. Later, thanks to the development of strong new ropes and the use of expensive sails of cotton or linen canvas, it was possible to simplify and reduce the rigging. This meant that the crew of each vessel could number as few as twenty-five people, and the length of their voyages without restocking could be extended. The ships were built with shallow drafts. That, in combination with their ability to sail close to the wind and the maneuverability provided by a rudder hung from an axle, made them the best boats to explore the African coast against the prevailing winds.

At the nautical academy Henry founded at Sagres, on the southwestern tip of Portugal, new methods of navigation and ways of using the stars were developed. As the caravels began to make their way south and the northern stars were lost beneath the horizon, navigation methods that read off the height of the polestar on an astrolabe to give latitude could not be used. Instead, tables were developed that enabled the sun to be used as a substitute. But of the greatest importance was that systematic information gathering and mapmaking were begun. The caravels were expected—and eventually required—to bring back detailed logs of their voyages, which enabled not only the route around Africa to be mapped but also the pattern of ocean currents.

Even so, the voyages attempted by the Portuguese were daunting. In 1410, nothing was known of Africa south of Cape Bojador (which lies just beyond the modern border of Morocco). From Cape Bojador onward, the shore was barren, backed by the great desert, and the winds and current were unfavorable for a return to Portugal. It was believed that no one who passed the cape would ever return. Henry's first great success was in his expeditions that discovered the islands of Madeira (1419) and the Azores (1431). These offshore islands enabled the crews to find safer routes home from North Africa by swinging out to sea and avoiding unfavorable currents. Even so, it took a further fifteen attempts for the first ships to pass Cape Bojador and move south toward Senegal.

Through these decades, Prince Henry had few supporters for his expensive voyages of exploration. But in 1441, by which time caravel design had advanced rapidly, one of his ships traveled far enough south to make contact with African cultures. The ship returned to Portugal with two Africans. It was the first direct contact by sea between Europe and black Africa and created enormous excitement well beyond the borders of Portugal.

The Age of Discovery began. In 1460 Henry died, with his last expedition having reached Cape Verde, still far short of the point where the African coastline turns east toward modern Nigeria. But before the end of the century Vasco da Gama had rounded southern Africa and reached India. Separate cultures came into contact with one another (sometimes with horrific consequences, as the slave trade took root) and international sea trade began. A turning point in history had been passed, thanks to the tiny, brilliantly engineered caravels and the new skills of navigation and mapmaking.

ALUN ANDERSON is the editor of *New Scientist*.

SAMUEL H. BARONDES
Organized Science

The great invention of the modern era is organized science—scientific societies and journals that foster the accumulation and dissemination of knowledge based on evidence rather than on authority or revelation. Before the invention of these organizations, the accumulation of scientific knowledge was slow, because there were no established venues for communication and criticism—essential processes that stimulate new ideas, refute the untenable, and provide a system of recognition and reward based on merit and true achievement. Although these organizations are becoming very large and impersonal, necessitating the proliferation of many subdivisions to allow for interactions on a human scale, they remain the essential social structures at the heart of science—the brilliant invention that makes possible the daily growth of knowledge that so many of us enjoy.

SAMUEL H. BARONDES, M.D., is the Jeanne and Sanford Robertson Professor and director of the Center for Neurobiology and Psychiatry at the University of California at San Francisco, and the author of *Molecules and Mental Illness* and *Mood Genes: Hunting for Origins of Mania and Depression.*

JOHN R. SEARLE

The Green Revolution

If by "invention" we mean technological advances as opposed to ideas, theories, and concepts, then there have been some good ones. One thinks of the printing press and the clock, for example. It is too early to say for sure, but my choice for the most important invention of the past two thousand years would be the set of agricultural techniques known collectively as the green revolution. This invention began in the 1960s and continues into the 1990s; indeed, it is now developing into something that may well come to be called the green-blue revolution, which is extending new agricultural techniques to the oceans.

The most important invention of all time occurred with the neolithic revolution, when humanity found ways to grow crops systematically and thus overcame both the instability and the mortal perils of the hunter-gatherer way of life. Hunter-gatherers could neither stay in one place long enough to develop a stable civilization nor count on being able to survive periods of drought and other forms of natural catastrophe. With the neolithic revolution, both of these problems were solved, and civilization became a real possibility.

However, the neolithic revolution brought problems of its own—in particular, the Malthusian problem. The growth of population was constantly threatening to outrun the growth of the food supply. For the foreseeable future, at least, this problem has been solved by the green revolution. The food supply has vastly outrun the increase in population. Nowadays if you read that there is a famine going on in some part of Africa or Asia, you know that it has been deliberately, politically, created. There is no international shortage of food. There is plenty of food to go around, and because of the green revolution there will be food to go around for a long time to come.

JOHN R. SEARLE is Mills Professor of Philosophy of Mind at the University of California at Berkeley. His books include *Intentionality: An Essay in the Philosophy of Mind; Minds, Brains, and Science; The Rediscovery of the Mind; The Construction of Social Reality; Mind, Language, and Society;* and *The Mystery of Consciousness.*

MARC D. HAUSER

The Electric Light and Aspirin

There is an old joke that goes something like this: A guy is taking a national poll on the most extraordinary invention of all time. During his travels, he finds himself in the Deep South and encounters a distinguished old gentleman rocking in his chair on the front porch of his house. The pollster approaches him and says, "Sir, if you don't mind, I would like to ask you a question for the poll I am conducting. I am interested in finding out what people consider to be the greatest invention of all time. Do you have an opinion on this?"

The old gentleman scratches his head and replies, "Well, I would have to say the thermos."

This baffles the pollster. "Sir, of the millions of responses I have collected, not one person has mentioned the thermos. Would you kindly tell me why that's your choice?"

"That's easy," says the old guy. "You see, the thermos keeps cold things cold and hot things hot. But how does it know?"

OK, I have two suggestions of my own. First, the electric light, born in 1828, about fifty years before Joseph Swan patented the incandescent lamp. Having lived in Africa, where one is often forced to read by firelight, I have found electricity to be a godsend. Moreover, once the incandescent lamp had been invented, it didn't take too long to come up with the flashlight, a handy device for those of us working in dark jungles.

My second suggestion for great inventions is the aspirin, invented in 1853—oddly enough, in France. Clearly, other medicines have been around, many of which serve comparable functions, but what a useful little pill for reducing headaches, body aches, and fevers! Among the Masai, headaches are treated with a mud compact of goat feces to the forehead. I prefer aspirin.

MARC D. HAUSER, a cognitive neuroscientist, is a professor in the Department of Psychology at Harvard University, where he is a fellow of the Mind, Brain, and Behavior Program; he is also the author of *The Evolution of Communication.*

JOHN D. BARROW
The Indo-Arab Counting System

The most important invention is the Indo-Arab counting system: 0, 1, 2, 3, 4, 5, 6, 7, 8, 9, with its positional information content (so that 111 means "one hundred plus one ten plus one unit"), its zero symbol, and the operator property that the addition of a zero to the righthand end of a string multiplies a number by the base value of 10. This system of counting and enumeration is now universal, and it lies at the foundation of all quantitative science, economics, and mathematics.

JOHN D. BARROW is a professor of mathematical sciences and the director of the Millennium Mathematics Project at the University of Cambridge, England. His books include *Pi in the Sky; Theories of Everything; The World Within the World; The Anthropic Cosmological Principle* (with Frank J. Tipler); *The Artful Universe: The Cosmic Source of Human Creativity; Impossibility: The Limits of Science and the Science of Limits;* and *Between Inner Space and Outer Space.*

LEON LEDERMAN
The Printing Press and the Thermos Bottle

If we suggest anything other than the printing press, Brockman will cancel our Christmas bonuses and the New Year's Eve turkey. So: the greatest invention in the past two thousand years is the printing press. Next is the Thermos bottle.

LEON LEDERMAN, the director emeritus of Fermi National Accelerator Laboratory, has received the Wolf Prize in physics (1982) and the Nobel Prize in physics (1988). In 1993 he was awarded the Enrico Fermi Prize by President Clinton. He is the author of several books, including (with David Schramm) *From Quarks to the Cosmos: Tools of Discovery* and (with Dick Teresi) *The God Particle: If the Universe Is the Answer, What Is the Question?*

RICHARD POTTS

Flying Machines

Over 4.6 billion years, the most important evolutionary inventions have been those that code, store, and use information in new ways: DNA, nervous systems, organic devices enabling the cultural transmission of information. In large perspective, the most important invention of the past two thousand years will likely be something related to computers—to electronic information being coded and handled outside living bodies. But I'm going with something whose impact, so far, is more apparent. The paleontologist in me wants to say something like "the discovery of time"—from inventions that have led to an intense sense of personal time to those that investigate the age of the universe or the human species. These inventions are perception-altering. But there's another invention with greater impact.

My vote is for flying machines, an invention that taps into the center of our mythologies. Prehistoric humans found ways of overcoming fire, water, and earth with the inventions of hearths, boats, and the wheel. The conquest of air did not begin until the twentieth century, with flying machines.

Aircraft have altered our perceptions in ways that were evolutionarily unpredictable. Many inventions change our lives but keep us in the prior range of human possibilities (or human nature). Firearms, as an example, mainly extended existing tendencies to bluff, subjugate, or kill in immediate, face-to-face situations. Aircraft, on the other hand, enabled the delivery of weapons, vastly destructive weapons, on an intercontinental scale—a scale unprecedented in evolutionary history. A flu virus that mutates in Kennedy Airport is spread around the world within a day or two. And so the history of disease has been altered by moving the month- or year-long dispersal of disease to a timescale of hours.

We now meet other people anywhere in the world in less than a day's travel. Thus things foreign and strange have become familiar. Ancient phobias, ancient biases, have been altered widely. The CNN culture of instantaneous worldwide information is an extension of this, but in my view the actual intermingling of people has been the more important, precedent-shattering development. Civil strife remains the worst where cultural and physical insularity reigns.

Finally, flying machines have meant a global alteration in how we allocate food and other resources. Humans are now bound together in a worldwide economy (resource exchange) driven by our interdependence. Two million years ago, the movement of such resources as food and stone tools was a development with extraordinary implications for human evolution. But even two thousand years ago no one could have foreseen just how far this process of resource exchange has gone today largely because of flying machines.

RICHARD POTTS is the director of the Human Origins Program at the National Museum of Natural History, Smithsonian Institution.

He is the author of *Early Hominid Activities at Olduvai* and *Humanity's Descent: The Consequences of Ecological Instability.*

PAOLO PIGNATELLI

The University

My choice is the university. Knowledge increases through synergy—through the spreading activation of millions of neurons, families of neurons, and neuron families distributed among thinking individuals. Universities, in bringing individuals with a common intellectual foundation into close enough proximity to allow for rich communication, cause jumps across metaphorical collective neurons—signals that then propagate through the society of neurons (à la Minsky) and create new knowledge. Universities are about expanding our universe. Had John asked us this question two thousand years ago, my choice would have been the library.

PAOLO PIGNATELLI owns and operates two Internet companies: www.TheCornerStore.com, a software retail company, and ContentPlace.com, a media company.

DOUGLAS RUSHKOFF

The Eraser

The eraser. As well as the delete key, correction fluid, the constitutional amendment, and all the other tools that let us go back and fix our mistakes.

Without our ability to return, erase, and try again, we would have no scientific model nor any way to evolve government, culture, or ethics. The eraser is our confessor, our absolver, and our time machine.

DOUGLAS RUSHKOFF is the author of *Coercion: Why We Listen to What "They" Say,* as well as *Cyberia, Media Virus, Playing the Future, The GenX Reader, Stoned Free,* and the novel *Ecstasy Club.* His books have been translated into sixteen languages, and his newspaper column is syndicated by the *New York Times.* He teaches virtual culture at New York University, delivers commentaries on NPR, and lectures around the world about the impact of media and technology on culture.

VIVIANA GUZMAN

Television

Why hasn't anyone mentioned television? Is it too obvious? I think it's the single most powerful and manipulative tool ever invented. It's today's most important source of information, and serves as a tremendous behavior-patterning device. Since its inception crime has risen, sex has increased, and the attendance at live performances has declined precipitously.

VIVIANA GUZMAN is a flutist and a graduate of the Juilliard School. Her classical album *Flute Fantasies* features the music of G. P. Telemann. She currently tours worldwide with her own World-Beat music group, heard on her album *Planet Flute*.

GARNISS CURTIS

Gutenberg's Press with Movable Type

My instantaneous response was Gutenberg's printing press with movable type. This knee-jerk response was followed by a pause and reflection. What is meant by "invention"? So, to the dictionary! Essentially, the word designates anything that did not exist previously, whether it be a mechanical device or art, literature, or music. Thus sobered, I reflected again.

In the 1930s, ten eighty-thousand-year-old skulls were found in Skhul Cave, at the foot of Mount Carmel in Israel; in size and shape they closely resemble the skulls of modern *Homo sapiens*. A similar skull, over ninety thousand years old, was found in a cave at Qafzeh, in Israel. That the braincases of these ancient hominids were the same as ours does not, of course, necessarily mean that their intelligence was the same as ours—although they were capable of making excellent stone tools.

We jump now to the Chevaux cave, in France, where wall paintings of animals extant in Europe at that time are beautifully depicted and have been dated at more than thirty thousand years old. Fifteen thousand years later, in the caves at Le Portel and Lascaux, our ancestors were making magnificent polychrome paintings of animals. Their stone tools, developed some five thousand years earlier, are comparable in technique and beauty to the much more recent tools of pre-Columbian America. Can anyone doubt that these Cro-Magnons, if magically brought into our present-day culture, could have learned to read and write, to philosophize, to do math at a high level, to learn chemistry and physics? (Let's not query those fundamentalists who still don't believe in evolution.)

Cuneiform writing began about five thousand years ago and quickly evolved. Twenty-five hundred years later, the Greeks were producing masterpieces of plays, literature, art, and architecture, and they were doing some wonderful things in mathematics and elementary observational science. The Romans carried on these traditions until Rome fell. Christianity came in and destroyed as much as it could of this great heritage in western Europe, including the great library in Alexandria, and thus began the Dark Ages in Europe. The gradual dissemination of knowledge, other than that contained in Christian literature (which didn't occur much faster!), was extremely slow. So, in the mid fourteen hundreds, along comes Gutenberg with his printing press and its movable type. Of course, almost the first thing he did was print a Bible or two, and they sold like hotcakes. Fixed (nonmovable) type had been around for a while, but the process wasn't much faster than copying by hand and was very costly, so the rapid dissemination of knowledge through printed books commenced with Gutenberg.

While the Dark Ages began to end about the year 1000, real progress wasn't made until the Renaissance—particularly with the proliferation of books via Gutenberg-type presses. As books were published, people became inspired to learn to read. Reading led to thinking about what had been read, to further publication, and to communication between people. The first "World Wide Web" had arrived. Anyone with a grain of sense can see what this has led to!

So, John, after the considerations outlined above, I still think the Gutenberg press with movable type is the greatest invention of the past two thousand years—or perhaps even of the last five thousand years, after cuneiform writing was invented.

GARNISS CURTIS is professor emeritus in the Department of Geology and Geophysics at the University of California at Berkeley, and the founder of the Berkeley Geochronology Center. A colleague of Louis Leakey, he determined the age (1.85 million years) of the famous Zinjanthropus fossil, a finding that rocked the anthropological world. His research in this field continues, and in 1994, with his colleague Carl Swisher, he redated *Homo erectus* in Java at 1.8 million years instead of the long-held eight hundred thousand years.

SUSAN BLACKMORE
The Contraceptive Pill

Birth control. (Or if you need it to be more specific, the pill.) Why? Because freedom from constant childbearing means that women can become as effective at spreading memes as men are. Memes are ideas, skills, habits, fashions, or stories that are spread by imitation. Like genes, they compete to get copied, and in the process construct our minds and culture. Women with fewer children can spread more memes, and this changes the kinds of memes that propagate effectively, including the memes of other inventions, and of science, arts, religion, and the media. In other words, the contraceptive pill is important because it has changed the whole of culture. Few single inventions have had this effect on the entire meme pool.

SUSAN BLACKMORE, a reader in psychology at the University of the West of England, in Bristol, is a columnist for the *Independent*. Her books include *Beyond the Body; Dying to Live: Near-Death Experiences; In Search of the Light: The Adventures of a Parapsychologist;* and *The Meme Machine*.

———————◆———————

PATRICK BATESON
The Harnessing of Electricity

As I sit at my computer writing this whimsy, I realize how much of my life is spent peering at its pale screen. So much of my working life has been transformed by the user-friendly software that is now available. As an inveterate reviser, when I write by hand I start to change my prose almost immediately after I have written something. Large chunks are crossed out, word orders are changed, sentences rearranged, paragraphs moved about. Before long the manuscript looks like a bird's nest. Producing a tidy typewritten copy is not at all easy after so many afterthoughts. The editing facilities of modern word-processing packages are so straightforward that manuscript birds' nests are a part of my past. The new technologies have been truly liberating. So my first thought was that the invention of friendly word processors was my candidate for this symposium. But wait a minute.

A good principle used by historians of technologies is to ask what had to be known in order for a particular development to have occurred. It is doubtful, for example, whether desktop computers of the power and flexibility we now have would have been possible without the invention of the silicon chip. This approach to emerging technologies produces a fan of necessary developments—or, more aptly, a root system branching outward as the historian moves backward in time. Some of these roots are undoubtedly more important than others, some certainly more enabling. Consider the computer on my desk again. It is inconceivable that such a machine would have been possible without electricity.

To be sure, Charles Babbage developed plans in the 1830s for what he called an analytical engine. His idea was that the machine would perform any arithmetical operation on the basis of instructions from punched cards, memory units in which to store numbers, sequential control, and most of the other basic elements of the present-day computer. The analytical engine was not built according to Babbage's specifications for another hundred and fifty years. Its mechanical components meant that it was bulky, and the modern outgrowth of a Babbage machine would exclude both my desk and me from the room in which it sat—and it would do a fraction of what my liberating machine does. So, my candidate for the greatest invention of the last two thousand years is the harnessing of electricity.

The first device that could store large amounts of electric charge was the Leyden jar, invented in 1745 by Pieter van Musschenbroek, a Leyden physicist. The jar was partially filled with water and contained a thick wire capable of storing a substantial amount of charge. One end of this wire protruded through the cork sealing the jar and was connected to a device that generated friction and static electricity. Soon after the invention, "electricians" were earning their living all over Europe killing animals with electric shock and devising other spectacles. In one demonstration in France, a wire made of iron connected a row of Carthusian monks; when a Leyden jar was discharged, the white-robed monks leapt simultaneously into the air. The frivolities led to thought. Thanks to Ben Franklin in the United States and Joseph Priestley in England, experiments and theorizing proceeded apace, and by the mid-nineteenth century the study of electricity had become a precise, quantitative science which paved the way for the technologies we now all take for granted.

We need electricity for keeping us cool in summer and warm in winter—though our ancestors would have been flabbergasted by the profligate way in which we do so. We use electricity for cooking much of our food and for freezing what we intend to eat later. We depend on it for transport, for communication, for entertainment, for running lives that bear no relation to the rising and setting of the sun. Of the major human appetites, only sex, it seems, is likely to be served by a power cut.

PATRICK BATESON is a professor of ethology at Cambridge University; the provost of King's College, Cambridge; and a fellow and biological secretary of the Royal Society of London. He was the director of the Sub-Department of Animal Behavior at Cambridge for ten years. He is the coauthor (with Paul Martin) of *Measuring Behavior* and *Design for a Life: How Behavior Develops*. He has also edited or co-edited several books, including *Mate Choice, The Development and Integration of Behavior, Behavioral Mechanisms in Evolutionary Perspective,* and the series Perspectives in Ethology.

CARL ZIMMER

Waterworks

I nominate waterworks, the system of plumbing and sewers that gets clean water to us and dirty water away from us. I'm hard pressed to think of any other single invention that has stopped so much disease and death. It may not inspire quite the intellectual awe as something like a quantum computer, but the sheer heft of the benefits it brings about so simply makes it all the more impressive. John Snow didn't need to sequence the *Vibrio cholerae* genome to stop people from dying in London in 1854—he didn't even know what *V. cholerae* was—but a pattern of deaths showed him that to stop a cholera outbreak all he needed to do was shut down a fouled well. Without waterworks, the crowded conditions of the modern world would be utterly insupportable, and you only have to go to a poor city without clean water to see this.

Another sign of the importance of an invention is the havoc it can wreak, and waterworks score here again. By cutting down on infant mortality, waterworks have helped to fuel the population explosion, and they have also allowed places like Las Vegas to suck the surrounding land dry.

I'd even go so far as to put the importance of the invention of waterworks on an evolutionary scale with things such as language. For hundreds of millions of years, life on land has been crafting new ways to extract and hold onto water. With plumbing, however, you don't go to the water—the water comes to you.

CARL ZIMMER is the author of *At the Water's Edge: Macroevolution and the Transformation of Life*. A former senior editor at *Discover*, he currently writes a column for *Natural History* and contributes articles to *National Geographic, Audubon,* and other magazines.

ROBERT SHAPIRO

Genetic Sequencing

Most of the inventions mentioned thus far affect the boundaries between human beings and the natural world. But the operations of the human body and the brain it contains support all of the experiences that make up our existence. Discoveries that will permit us ultimately to take charge of these operations and shape them to our desires surely deserve nomination as the most important of the last two millennia.

These insights flow broadly from the area of science that is now called molecular biology, but if I had to single out the most important invention that makes the process possible I would select genetic sequencing. The new techniques developed by Fred Sanger at Cambridge and Walter Gilbert at Harvard in the mid-1970s have allowed us to read out rapidly the specific information stored in our genes and those of all other living creatures on earth.

The new methods stimulated a burst of scientific energy that will culminate in the next decade, when the sequence of about 3 billion characters of DNA encoding a typical human being will be fully deciphered by the Human Genome Project. In subsequent explorations, we shall determine how individuals differ in their heredity and how this information is expressed to produce the human body.

Thus far the effects of sequencing have largely impacted us through such media-worthy events as the identification of the stain on Monica Lewinsky's dress, validation of the identity of the

Romanov bones, refutation of the claim of Anna Anderson to be Anastasia, and confirmation of Thomas Jefferson's affair with Sally Hemings. Much, much more is yet to come. The completion of the Human Genome Project will provide us with an understanding, at the molecular level, of human hereditary disease; much has already been learned about Huntington's disease, cystic fibrosis, and others. Further, by the application of other tools from modern molecular biology, we will be able to do something about these afflictions in the near future. They will be treated and (if society permits it) corrected at the genetic level. Beyond that, we shall come to understand and perhaps control many unfortunate aspects of the human condition which have until now been taken for granted—everything from baldness to aging. Ultimately we may elect to rewrite our genetic text, changing ourselves and the way in which we experience the universe.

I will also suggest that any poll taken now would not do justice to this invention, as most of its consequences still lie ahead of us. Perhaps we should schedule another poll for the year 3999, when the passage of two additional millennia will have given us more perspective.

ROBERT SHAPIRO, a professor of chemistry at New York University, is the author of *Origins: A Skeptic's Guide to the Creation of Life on Earth; The Human Blueprint;* and *Planetary Dreams: The Quest to Discover Life Beyond Earth.*

HOWARD GARDNER

Classical Music

Agood question! My perhaps eccentric but nonetheless heart-felt nomination is Western classical music, as epitomized in the compositions of Bach, Beethoven, Brahms, and, above all, Mozart. Music is a free invention of the human spirit, less dependent upon physical or physiological inventions than most other contrivances. Musical compositions in the Western tradition represent an incredible cerebral achievement, one that is not only appreciated but also imitated or elaborated upon wherever it travels. Most inventions, from nuclear energy to antibiotics, can be used for good or ill. Classical music has probably given more pleasure to more individuals, with less negative fallout, than any other human artifact. Finally, while no one can compose like Mozart and few can play like Heifetz or Casals, anyone who works at it can perform in a credible way—and, courtesy of software, even those of us unable to play an instrument or create a score can now add our own fragments to an ever expanding canon.

I've picked Western classical music in part because it has such deep and personal meaning for me. A more general answer I could have given was "the scholarly disciplines" or simply "the disciplines." Those of us who are educated take disciplines like history, mathematics, science, and the several arts for granted. We all too often forget that these disciplines were developed—symbol by symbol, concept by concept, method by method, so to speak—over many centuries by countless reflective individuals. The disciplines are what separate us from the barbarians. Our medieval ancestors understood this: young scholars had to master the trivium and the quadrivium, which included music along with arithmetic, geometry, and astronomy. What an irony that the quadrivium fell out of fashion at the very time that classical music was coming into its own.

HOWARD GARDNER is a professor of education at Harvard University. His numerous books include *The Mind's New Science: A History of the Cognitive Revolution; Frames of Mind; Leading Minds; Creating Minds; Extraordinary Minds; The Disciplined Mind;* and *Intelligence Reframed: Multiple Intelligences for the 21st Century.*

———————◆———————

ROGER C. SCHANK

The Internet

The Internet is by far the most important invention of the past two thousand years. Of course, it relies on numerous other inventions—chips, networking, CRTs, telephones, electricity—but its influence is far greater than the sum of its parts. There are two reasons why it might not be an obvious choice for this distinction. First, it has become so prevalent in our lives that many people actually fail to notice it. Secondly, its power has not yet begun to fully manifest itself.

Information delivery methods already radically affect almost every aspect of how we live. We still go to schools, offices, the post office, places of entertainment, or shopping malls; but when we don't have to walk to town to find out what's going on, or to shop, or to learn, why should we go to town at all? Schools will soon transform themselves, as we are able to build better courses on the Internet than could possibly be delivered in classrooms. Here's an example of what I mean. We now have the technology necessary to build courses with the best physicists (or mathematicians or biologists) available on-line, not only to deliver learn-by-doing simulations that allow students to try things out, do experiments, and solve real problems, but also to answer student questions and offer advice at just the moment when the student needs help. As courses like this are built in all fields, and students can use the Internet to take the courses when they're interested in them and want to learn from them, colleges and universities as we now know them will become obsolete, with their only enticement being football and fraternity parties.

And this sort of radical change won't be limited to schools. Shopping malls aren't gone yet, but they will be. For instance, fewer and fewer people are going to stores to buy CDs. Using the Internet, they can listen to samples of music, then simply click a button on a website to arrange for delivery of the CD they want, all the while sitting at home. Any object that needn't be felt or perused prior to purchase will find no better delivery method than the Internet. Newspapers? Not dead yet, but they will be. Pick an aspect of the way we live today, and it will change radically in the coming years because of the Internet. Life (and human interaction) in fifty years will be so different that we will hardly recognize the social structures that will evolve. I don't know whether we'll be happier, but at least we'll be better informed.

ROGER C. SCHANK, a computer scientist and cognitive psychologist, is the director of the Institute for the Learning Sciences at Northwestern University, where he is John Evans Professor of Computer Science, Psychology, and Education and Social Policy. His books include *The Creative Attitude: Learning to Ask and Answer the Right Questions; Tell Me a Story; Engines for Education;* and *Virtual Learning.*

---◆---

RANDOLPH NESSE

Printing

It seems to me, as it will no doubt to many others, that the printing press has changed the world more than any other invention in the past two millennia. But why has such a simple technology had such a huge influence? And why, after five hundred years, has no one invented a superior replacement?

I suspect it is because text is special. It has a unique relationship to the design of the human mind and has played a central role in developing our minds and cultures. It is the third wave of the biggest innovation—the one that started with the coevolution of language, thought, and speech.

Speech allows us to share and compare internal models of the external world, an ability that gave the human species a huge selective advantage. But acoustic vibrations are ephemeral, fading in moments into questions about who said what, when.

Writing, the second wave, was like a blast of supercooled air that froze words in midflight and smacked them onto a stone tablet or a scroll, where they could be examined by anyone, anywhere, anytime. Writing made possible law, contracts, history, narratives, poetry—to say nothing of sacred texts, with their overwhelming influence.

Printing, the third wave, transformed writing into the first mass medium, and the world has never been the same since. In the half-century that followed Gutenberg's 1455 Bible, over a thousand publishers printed over a million books. Suddenly it was worthwhile, and soon essential, even for ordinary people to learn to read. Nowadays people whose brains have trouble with this trick are at a severe disadvantage, while some with unusual verbal facility can make a living just by arranging words on paper.

Is text merely a temporary expedient, necessitated by our previous inability to record and transmit speech and images? We will soon see. In just a few years, sensors, storage, and bandwidth will be so inexpensive that many people will be unconstrained by technical limitations. This affords a fine opportunity to make bold predictions that can be completely and embarrassingly wrong—as wrong as the predictions that e-mail would never catch on. In that spirit, I predict that voice and video attachments to e-mail— namely, v-mail and vid-mail—will be the next big thing, and that they will create all manner of consternation. At first they will be hailed as more personal and more natural media, thanks to the increased content carried by intonation and exclamations. But soon, I predict, the usual human strivings will give rise to problems.

Many people who previously were forgiven for "liking to hear themselves talk" will be revealed as actually needing to have others hear them talk. Some, especially bosses, will send long soliloquies to hundreds of other people, in the expectation that they will be listened to in full. The wonderful veil of privacy in which a reader considers a text will be rent. We won't be able to get away with jumping around and skipping whole paragraphs in v-mail and vid-mail, as we can in e-mail. Time and attention will be revealed as the valuable resources they are. Seeing our words in print gives us the satisfying illusion that we are communicating with millions, but most of us (except those bosses) will soon realize that hardly anyone is interested in listening to our pronouncements. Some people will post electronic notices equivalent to the one a friend has on his answering machine: "Leave a message, but please KEEP IT BRIEF."

To cope with the social dilemmas spawned by this technology, we will turn, of course, to still more technology. V-mail will be transformed automatically into text, so that we will have a choice of media. Which will we choose? It will depend. For emotional endearments and many narratives, v-mail and vid-mail will be preferred. For simple facts and subtle ideas, however, I think we will choose the glorious invention that created a new kind of privacy— text. That is, until our brains are changed by the selective forces unleashed by the new media.

RANDOLPH NESSE, M.D., is a professor of psychiatry and director of the Evolution and Human Adaptation Program of the Institute for Social Research at the University of Michigan, and coauthor (with George C. Williams) of *Why We Get Sick: The New Science of Darwinian Medicine*.

RON COOPER
Distillation

I am surprised that no one has mentioned distillation, the great alchemical invention of transformation in the search to understand the essence of existence.

Alchemy appears to have started in ancient Egypt. (*Al-khem* means "the art of Egypt" in Arabic.) With Islam, it spread across northern Africa, and it traveled to mainland Europe with the Moorish invasion of Andalucia in the tenth century. Alchemy tries to make sense of the world by, among other things, working with the elements to transform matter and strip away the extraneous and capture its purest essence. Some suggest that alchemy's found-ing father was the Egyptian god Thoth (in Greek, Hermes). Both are symbols of mystical knowledge, rebirth, and transformation.

To find the first evidence of distillation of spirits, you have to go to fourth-century China, where the alchemist Ko Hung wrote about the transformation of cinnabar in mercury as being "like wine that has been fermented once: it cannot be compared with the pure clear wine that has been fermented nine times." Is he talk-ing about distillation? It seems possible; how do you ferment a wine nine times unless you distill it? By then the Alexandrian Greeks had discovered that by boiling you could transform one ob-ject into another. Pliny writes about distillation being used to ex-tract turpentine from resin, while Aristotle recounts how seawater can be turned into drinking water. Aside from alchemy's being the basis of modern science and industry, the transformation of hu-man beings brought on by the imbibing of distilled spirits is of great interest to me.

RON COOPER is an artist whose work has been featured in group shows and solo exhibitions around the world and is in many prominent public collections, including the Whitney Museum of American Art, the Guggenheim Museum, the Los Angeles County Museum of Art, the Chicago Art Institute, Amsterdam's Stedlijk Museum, and the Bibliothèque Nationale in Paris. He is also the founder and president of Del Maguey, Single Village Mezcal, a company dedicated to producing intoxicating elixirs distilled from the maguey that tap into the mythical past.

DAVID BUSS

Television in its Effects on Mating Patterns

It is difficult to determine "the most important invention" in the absence of answers to the prior question: "Important with respect to what?" One criterion for "the most important invention" is whether, and to what degree, it has altered patterns of human mating. Changes in mating can affect the subsequent evolutionary course of the entire species, with cascading consequences for virtually every aspect of human life. Who mates with whom has been the subject of intense interest among scientists, ranging from biologists and geneticists to psychologists and sociologists, because of this range of impact. Most people marry at some point in their lives, and these marriages affect social trends, such as the distribution of wealth. Universities and colleges affect mating by placing similar individuals into close proximity, thereby promoting assortative mating. Patterns of selective mating can affect the genetic structure of the population in subsequent generations: those who are shunned fail to reproduce; those who possess desired qualities preferentially pass on their genes.

Although many inventions have altered human mating over the past two thousand years, television and related media must rank among the most important. Television has changed status and prestige criteria, creating instant celebrities and hastening the downfall of leaders. For example, an obscure waiter or waitress can become an overnight sensation because of media exposure—an event unlikely to have occurred previously over the long course of human evolutionary history. With fame comes sexual access to more mates and more desirable mates, in numbers orders of magnitude in excess of those experienced by our ancestors.

The visual media have likely increased the importance of physical appearance. The importance of appearance in a potential spouse has been documented roughly every decade since the 1930s in America. Its importance steadily rises over time, being valued roughly twice as much today as it was before television. The importance attached to physical appearance, in turn, has accelerated the intensity of intrasexual mate competition, particularly among women. The skyrocketing rates of eating disorders in America, spreading to European countries, likely originates in runaway intrasexual competition caused by the increasing importance that people place on appearance. Heightened same-sex competition, in turn, has detrimental effects on self-esteem and perhaps even on the quality of existing relationships. Research shows that women's exposure to a barrage of physically attractive female models lowers their self-esteem. Men's exposure to attractive female models decreases the commitment they feel toward their regular partner. The consequence may be an overall decrease in psychological well-being and a pervasive sense of dissatisfaction with existing relationships.

These and related phenomena have acutely transformed the nature of sexuality and mating. We can't or won't go back to a time before media images; they mesmerize us, exploiting our evolved psychological mechanisms. And since visual media may alter who mates with whom, they may forever change the evolutionary course of our species.

DAVID BUSS is a professor of psychology at the University of Texas at Austin. He is the author of *The Evolution of Desire: Strategies of Human Mating* and *Evolutionary Psychology: The New Science of the Mind*.

DAN SPERBER

The Computer and the Atomic Bomb

The two most important inventions in the past two thousand years are the computer and the atomic bomb. The computer will bring about the greatest change to human life since the neolithic revolution, unless the bomb destroys human life altogether.

DAN SPERBER is a cognitive scientist at the Centre National de la Recherche Scientifique, in Paris. He is the author of *Rethinking Symbolism; On Anthropological Knowledge; Relevance: Communication and Cognition* (with Deirdre Wilson); and *Explaining Culture: A Naturalistic Approach.*

MARIA LEPOWSKY

The Pill, the Gun, and Hydraulic Engineering

I've been pondering your bimillennial question, and I'd like to cheat a bit by giving several answers.

I, too, cast a vote for the oral contraceptive pill. It is revolutionary for two reasons. First, it represents a tremendous leap in the effectiveness of attempts to control human fertility—attempts found in every known culture and likely dating back more than a hundred millennia. The pill and subsequent devices presage a revolution in the lives of women from puberty to menopause everywhere in the world, enabling half the human population to control their adult lives by controlling their own fertility.

In addition, these devices may well save the planet Earth from the ongoing disaster of human overpopulation, with its present and future dire global consequences of mass poverty, pandemics, warfare, violent confrontations over scarce resources, environmental degradation, and wholesale species extinctions.

My next vote for the most important technology of the last two thousand years goes to the gun—or, more precisely, to a series of European inventions of more efficient killing technologies. The ship-mounted cannon, the Spanish trabuco, and the British Snider rifle—to mention just a few weapons from recent centuries—in the hands of members of authoritarian societies whose populations had exceeded the carrying capacities of their homelands effected the European conquest of large portions of the planet's landmass, resources, and human populations. Bent on acquiring new territories, Europeans crossed the Atlantic and Pacific Oceans on ships built according to the most advanced maritime technologies of their eras. The momentous consequences of the European conquest will continue to play themselves out in every sphere of human life around the globe over the next millennium.

My final vote goes to the revolutionary improvements in hydraulic engineering in the late nineteenth century, which solved what had for millennia been the single greatest problem of urban life: how to bring clean water in and human waste out of a large, nucleated settlement. While the Roman waterworks were brilliantly designed (and their epoch crosses the bimillennial cutoff point of this exercise), improvements in sanitation made only a century or so before the present led, in industrial societies like Britain and the United States, to a dramatic drop in the death rate from infectious diseases transmitted by fecal contamination of drinking water. These advances in hydraulic engineering have extended human life spans even more than the subsequent discovery of antibiotics.

This technology has diffused only slowly around the globe as it encounters barriers created by unequal distributions of wealth and power. Even so, ironically, our resulting increased longevity, and the increase in fertility that declining mortality rates confer when they are unchecked by other variables, contribute hugely to the ongoing crisis of human overpopulation. This makes the wide availability of advanced contraceptive technology, invented two generations later, all the more critical for the survival and well-being of our species and of the entire planet.

MARIA LEPOWSKY is a professor in the Department of Anthropology at the University of Wisconsin, in Madison. She is the author of *Fruit of the Motherland: Gender in an Egalitarian Society.*

ROBERT R. PROVINE

Universal Schooling

Instead of suggesting a device, I nominate the educational process essential for a high velocity of inventiveness, the evolution of a technological society, and the spread of culture. While schools for the elite have existed since antiquity, the recognition of childhood as a unique time of life, with special schooling, social, and emotional needs and a different standard of justice, is relatively recent and is associated with Rousseau, Freud, and Piaget.

The discovery that children are not miniature adults led to a more humane society and to the tailoring of educational programs to the developmental stage of the student. Universal schooling (and even the modern university) were born of this increased appreciation of the special needs of children—and also out of necessity, since the industrial revolution needed trained workers, scientists, and engineers. The complexity of modern technology and the associated acceleration of innovation demand a critical mass of creative minds and hands which cannot be provided by occasional virtuosi toiling in solitude.

ROBERT R. PROVINE is a professor of psychology and neuroscience at the University of Maryland, Baltimore County.

———◆———

DUNCAN STEEL

The Thirty-Three-Year English Protestant Calendar

Summary answer: The nonimplemented 33-year English Protestant calendar.

The outcome of my mental perambulations on this question is that all the technological products of recent and latter years would have been invented sooner or later anyway, and in any case are mere applications of ideas. An idea may be important, even though it does not directly lead to a physical invention; an idea itself I count as being an invention in the current context. Further, how we got to where we are is the result of many important ideas producing bifurcations in history. One could make a case for the more ancient branching points being the more fundamental. (If

Alexander the Great, Charlemagne, and William the Conqueror had never lived, then neither would Hitler.) But that form of reasoning leads to a reductio ad absurdum. I choose to ask, "How did we get to where we are now?"—that is, with the U.S.A. being the powerhouse of most of the rest of the world. Economically, scientifically, technologically, mythologically, culturally (at least in popular culture), idealistically. Thus, the branching point I look to is that which made the U.S.A. a reality. I do not mean the Declaration of Independence; I mean, What made the English first go and settle the Atlantic seaboard of North America?

The answer I will give is not original to me but was suggested to me by Simon Cassidy, a British mathematician who lives in California.

Here is the story. When the Catholic Church, under Pope Gregory XIII, brought in the reformed calendar in 1582, it decided to use a second-best solution to the problem. To understand this, you must understand that all Christian calendar matters hinge on the question of the Easter computus, and that depends on the time of the spring equinox, which is ecclesiastically defined to be March 21 regardless of the real position of the sun in the sky; similarly, the ecclesiastical moon used in the Easter computus pays scant regard to the phase of the real moon in the sky. Astronomically speaking, the equinox on the Gregorian calendar shifts between March 19 and 21, over a 53-hour span, during the entire 400-year leap year cycle. That is, the Gregorian leap year rule (no leap day in a year number divisible by 100 but not 400) produces 97 extra days over 400 years; these days come close to keeping the average year length close to the desired value, but they cannot do so perfectly. Between 1901 and 2099 there is a leap every fourth year (2000 is not dropped, and has 366 days), with the result that there is a gradual drift in the instant of the equinox, the full range being that 53 hours. This leads to the astronomical equinox occurring on one of three dates, the long cycle time (400 years) producing this wander. The church rule simply ignores this and stipulates March 21 as the perpetual date instead.

By far preferable from a religious perspective would be a calendar that confined the equinox to one day, necessitating a shorter cycle. Even as far back as A.D. 1079, Omar Khayyám showed that 8 leap years in a 33-year cycle provided an excellent approximation to the year as measured between spring equinoxes. This 8/33 year cycle is better in two ways: it is closer to the real average year length and it keeps the equinox within a 24-hour spread. Gregory XIII's advisers knew this but instead recommended the inferior 97/400 leap year system we still use—perhaps thinking that the Protestants did not know of the better 8/33 concept.

But in England, they did. John Dee and others (Thomas Harriot and Walter Raleigh among them) had secretly come up with a plan to implement a "Perfect Christian Calendar" using the 33-year cycle—the traditional lifetime of Christ. In that span there are seven 4-year cycles, each producing a wander by the equinox of a little less than 18 hours. The problem is the one 5-year cycle (years 29 through 33 inclusive), during which the equinox steps forward by 5 hours and 49 minutes in each of four jumps before the following leap year with its extra day pulls it back by precisely 24 hours. The full amplitude of the movement is 23 hours and 16 minutes. One might expect that anywhere within an 11-degree band of longitude (corresponding to 44 minutes of time), one could keep the equinox on a single date using the 33-year leap cycle, but there are other influences. Some jitter in the equinox time occurs. Moreover, the question of whether one is using the real sun or the fictitious "mean sun" (hence "Greenwich Mean Time") is critical, since timekeeping deviates during the year by up to 16 minutes from the time indicated by the sun in the sky, because of the tilt of the earth's spin axis and the noncircularity of the earth's orbit. This narrows down the acceptable prime meridian for use in this calendrical sense. To get the equinox to remain on one calendar day throughout the 33-year cycle, one has to choose the meridian carefully. In the late sixteenth century, 77 degrees west was the indicated location—"God's Longitude." (The slowing of the earth's rotation under tidal drag—which makes the insertion of leap seconds necessary—and other minor variations have since shifted it to about 75 degrees west.)

In the 1580s, the settled areas along the 77th meridian (the Caribbean, Peru, etc.) were under Spanish—hence Catholic—control. In order to grab part of God's Longitude and found a New Albion, enabling them to introduce the rival Perfect Christian Calendar, England mounted various expeditions which historians have since misinterpreted, thinking them to be merely quests for new lands and new products. The real motivation was quite different: to seize a piece of the calendrical meridian in order to establish a better calendar than the Gregorian, a weapon with which to convert other wavering Christian states in Europe to Protestantism. In 1585, after a preliminary expedition the previous year, English colonists were dispatched to Roanoke Island, off the coast of what is now North Carolina. The so-called Lost Colony was a bizarre place from which to start the colonization of the New World but an excellent site from which to make astronomical observations to fix the longitude and thus decide how far inland New Albion should be. Similarly, in 1607 the choice of Jamestown Island, sixty miles up the James River, seems obtuse from the settlement perspective: why not out on Chesapeake Bay and away from the attacks of the local Algonquins? But the choices make sense when you consider the paramount need to grab a piece of God's Longitude. From the foothold the English managed to gain, Old Virginny grew, and later other colonizers came to New England, and New Amsterdam was seized from the

Dutch. But these later utilitarian developments do not reflect the original purpose of the English expeditions to Roanoke and Jamestown Islands—any more than that the Eiffel Tower was built to provide a mount for the many radio antennas which now festoon its apex.

After the fact, the English did not reveal their prime motivation for Raleigh's American adventures and the huge investment in the ill-starred Jamestown colony, and all of this is yet to be properly teased out. But if the English had never invented their nonimplemented 33-year Protestant calendar, then the U.S.A. as it is would not exist, and all of the scientific, technological, and cultural developments of the world over the past couple of centuries would be quite different. In view of this, I nominate that calendar, due to John Dee, as The Most Important Invention of the Past Two Thousand Years.

DUNCAN STEEL, who currently lives in England, conducts research on asteroids, comets, and meteors and their influence upon the terrestrial environment. He is also deeply interested in the ways in which various astronomical considerations have affected the tortuous development of civilization. Steel is author of *Rogue Asteroids and Doomsday Comets*, *Eclipse*, and *Marking Time*.

PETER TALLACK

The Stirrup and the Horse Collar

Mundane though it may seem, the horse has given rise to two low-tech inventions of inestimable importance in the development of almost all the great civilizations of the world: the stirrup and the horse collar.

The stirrup was unknown until the first centuries A.D., but once its use had become widespread, mounted soldiers could wield the lance and bowmen could shoot from the saddle, with very little training. Not only did this allow unskilled peasant farmers to enlist as cavalry, but protective armor could also increase in complexity. This in turn led to the breeding of much larger horses that could support the knight and his weapons. By medieval times, the heavy horse was taking over from the cattle for plowing and traction in northern Europe, ushering in what Juliet Clutton-Brock, in her 1992 book *Horse Power*, has described as the Age of the Horse. She points out that radical changes arose not simply because of the spread of the stirrup but also because of the invention of the horse collar, which was used in place of the neck strap as a method of harnessing, thus enabling horses to pull loads with greater efficiency.

In his influential 1962 study *Medieval Technology and Social Change*, Lynn White has argued that owing to these changes the horse became the essential pivot of civilization: the basis of long-distance transport, agriculture, industry, warfare, hunting, chivalry—indeed, the whole structure of feudal society. One of his more provocative suggestions is that many peasant farmers who had previously lived in small hamlets close to their fields began to group in villages of two or three hundred families. And, of course, armies of mounted knights could travel across the globe, engaging in battles that would alter the fate of nations.

PETER TALLACK, the former Book Review and Commentary editor of *Nature*, is the publishing director of science at Weidenfeld & Nicolson, London.

JOHN C. BAEZ

Social Structures that Enable Inventions

Here is my reply to your fiendish question:

How can we possibly pick the most important invention in the past two thousand years? The real biggies—language, fire, stone tools, agriculture, art—came much earlier. And in the last two millennia our world has seen so many inventions that it's hard to think of one that stands above all the rest. The printing press? The computer? The A-bomb? After a bit of this, one is tempted to give a smart-aleck reply and back it up with the semblance of earnest reasoning. "Thousand Island dressing! Because. . . ."

Well, if inventions are important, surely it was even more important to invent the social structures that guaranteed a steady flow of new inventions. I've heard it said that Edison was the first to turn invention into a business. Every day he would walk into his lab and ask, "OK, what can we invent today?" But the groundwork was laid earlier. Perhaps the invention of a patent office was the key step? Or further back, Francis Bacon's *New Atlantis* (1627), which envisioned the technoparadise we are now all so busy trying to build?

Once you get systematic about inventing everything you can possibly think of, individual inventions lose their importance.

JOHN C. BAEZ is a mathematical physicist at the University of California, Riverside. He is the coauthor (with Irving Segal and

Zhengfang Zhou) of *Introduction to Algebraic and Constructive Quantum Field Theory* (Princeton Series in Physics) and (with Javier Muniain) of *Gauge Fields, Knots, and Gravity.* He also writes a regular column entitled "This Week's Finds in Mathematical Physics," available on the Internet at http://math.ucr.edu/home/baez, and he enjoys answering physics questions on the usenet newsgroup sci.physics.research.

TERRENCE J. SEJNOWSKI
The Digital Bit

Technological advances in communication—from clay tablets to papyrus to movable type—have had a shaping influence on society, and these advances are accelerating. Almost overnight, the accumulated knowledge of the world is crystallizing into a distributed digital archive.

Images and music, as well as text, have merged into a universal currency of information, the digital bit, which is my choice for the greatest discovery of the last two millennia. Unlike other forms of archival storage, bits are forever. Clay breaks, papyrus crumbles, and paintings darken, but the information in a digital document is independent of the medium that is used to store it, and can be perfectly replicated.

In the next millennium, this digital archive will continue to expand, in ways we cannot yet imagine, greatly enhancing what a single human can accomplish in a lifetime and what our culture can collectively discover about the world and ourselves.

TERRENCE J. SEJNOWSKI, a pioneer in computational neurobiology, is regarded by many as one of the world's foremost theoretical brain scientists. He is an investigator with the Howard Hughes Medical Institute, the director of the Computational Neurobiology Lab at the Salk Institute, and a coauthor (with Patricia Churchland) of *The Computational Brain*.

◆

NICHOLAS HUMPHREY
Reading Glasses

The most important invention has been reading glasses. They have effectively doubled the active life of everyone who reads or does fine work—and prevented the world from being ruled by people under forty.

NICHOLAS HUMPHREY is a theoretical psychologist at the Center for Philosophy of Natural and Social Sciences, London School of Economics, and the author of *Consciousness Regained; The Inner Eye; A History of the Mind;* and *Leaps of Faith: Science, Miracles, and the Search for Supernatural Consolation*.

◆

CLIFFORD PICKOVER

Papermaking

In A.D. 105, Ts'ai Lun presented samples of his paper to the Chinese emperor Ho Ti. Ts'ai Lun was a member of the Chinese imperial court, and I consider his early form of paper to be humanity's most important invention—and the progenitor of the Internet. Although recent archaeological evidence places the actual invention of papermaking two hundred years earlier, and thus beyond the confines of your time frame, Ts'ai Lun played an important role in developing a material that revolutionized his country. From China, papermaking moved to Korea and Japan. Chinese papermakers also spread their handiwork into Central Asia and Persia, whence traders introduced paper to India.

Today's Internet evolved from the tiny seed planted by Ts'ai Lun. Both paper and the Internet break the barriers of time and distance and permit unprecedented growth and opportunity. In the next decade, communities formed by ideas will be as strong as those formed by geography. The Internet will dissolve away nations as we know them today. Humanity becomes a single hive mind, with a group intelligence, as geography becomes putty in the hands of the Internet sculptor.

Chaos theory teaches us that even our smallest actions have amplified effects. Now, more than ever before, this is apparent. Whenever I am lonely at night, I look at a large map depicting sixty-one thousand Internet routers spread throughout the world. I imagine sending out a spark, an idea, and a colleague from another country echoing that idea to his colleagues, over and over again, until the electronic chatter resembles the chanting of monks. I agree with the spiritualist Jane Roberts, who once wrote, "You are so part of the world that your slightest action contributes to its reality. Your breath changes the atmosphere. Your encounters with others alter the fabrics of their lives, and the lives of those who come in contact with them."

CLIFFORD PICKOVER is a research staff member at IBM's T. J. Watson Research Center, in Yorktown Heights, New York. He is the holder of more than a dozen patents relating to computer interfaces, and he has written some twenty books on a broad range of topics, including black holes, time travel, computer art, and the possibility of alien life. Pickover's primary interest is in finding new ways to expand creativity by melding art, science, mathematics, and other seemingly disparate areas of human endeavor. His Internet website has attracted nearly two hundred thousand visitors.

FREEMAN DYSON
Hay

This is a good question. My suggestion is not original. I don't remember who gave me the idea, but it was probably Lynn White, with Murray Gell-Mann as intermediary.

The most important invention of the last two thousand years was hay. In the classical world of Greece and Rome, and in all earlier times, there was no hay. Civilization could exist only in warm climates, where horses could continue to graze through the winter. Without grass in winter, you could not have horses, and without horses you could not have urban civilization. Sometime during the so-called Dark Ages, some unknown genius invented hay, forests were turned into meadows, hay was reaped and stored, and civilization moved north over the Alps. So hay gave birth to Vienna and Paris and London and Berlin, and later to Moscow and New York.

FREEMAN DYSON is a professor of physics at the Institute for Advanced Study, in Princeton, New Jersey. His professional interests are in mathematics and astronomy. Among his many books are *Disturbing the Universe, Infinite in All Directions, From Eros to Gaia*, and *Imagined Worlds*.

DANIEL C. DENNETT
The Battery

The battery, the first major portable energy packet in the last few billion years.

When simple prokaryotes acquired mitochondria several billion years ago, these amazingly efficient portable energy devices opened up Design Space to countless species of multicellular life. Many metazoa developed complex nervous systems, which gave the planet eyes and ears for the first time, expanding the epistemic horizons of life by many orders of magnitude. The modest battery and its sophisticated fuel-cell descendants, by providing energy for autonomous, free-ranging, unplugged artifacts of dazzling variety, is already beginning to provide a similarly revolutionary cascade of developments.

Politically, the transistor radio and the cell phone are proving to be the most potent weapons against totalitarianism ever invented, since they destroy all hope of the centralized control of information. By providing all of us with autonomous prosthetic extensions of our senses (think of how camcorders are revolutionizing scientific data-gathering possibilities, for instance), batteries enable fundamental improvements in the epistemological architecture of our species.

The explosion of science and technology that may eventually permit us to colonize space—or save our planet from a fatal collision—depends on our ability to store and extract electrical power ubiquitously. Electricity powers all the sensors (cameras, microphones, accelerometers, radio receivers, collision detectors, etc.) and effectors (motors, solenoids, switches) of our robotic artifacts, and even if these brainchildren of ours can be made to live off sunlight (or plutonium), they will need batteries to store that energy until it is required. Sol Spiegelman once said, "The nucleic acids invented human beings in order to be able to reproduce themselves even on the Moon." Human beings in turn invented batteries to power that quest.

Batteries are still no match for the mitochondrial ATP system—a healthy person with a backpack can climb over mountains for a week without refueling, something no robot could come close to doing—but they open up a new and different cornucopia of competences.

DANIEL C. DENNETT, a philosopher, is the director of the Center for Cognitive Studies and Distinguished Arts and Sciences Professor at Tufts University. He is the author of *Darwin's Dangerous Idea*, *Consciousness Explained*, *Kinds of Minds*, and *Brainchildren*, and a coauthor (with Douglas Hofstadter) of *The Mind's I*.

LAWRENCE M. KRAUSS

The Programmable Computer

If I take the word "important" to suggest an invention that will have the greatest impact on the next two thousand years (after all, it is the future that counts, not the past!), then the invention of the programmable computer seems to me to be the most important invention of the last two thousand years. (In my list of possibilities, I am not including ideas and concepts, since I don't think they qualify as inventions—and I suspect that the intent of the question is to explore technology, not ideas.) While the printing press certainly revolutionized the world in its time, computers will govern everything we do in the next twenty centuries. The development of artificial intelligence will be profound, quantum computers may actually be built, and I am sympathetic to the idea I first heard expressed by my friend Frank Wilczek—that computers are the next phase of human evolution. Once self-aware, self-programmable computers become a reality, I have a hard time seeing how humans can keep up without in some way integrating them into their own development.

We are hardwired by biology, and it is currently difficult to substantially alter the way we work and yet remain biologically viable. However, once computers become self-aware and self-programmable, radical improvements in their function will most likely occur at an exponential rate. We could integrate these improvements into our own development in one of several ways: (a) We could adapt our own biology to incorporate cybernetics directly into our functioning (the "Borg route"); (b) We could utilize the knowledge obtained by computers to improve our biological capabilities; (c) We could simply transfer our legacy of knowledge to computers, then sit back and let them lead the way; (d) Or, most probably, developments will occur in some way I haven't thought of.

The only other invention that may come close to the computer in importance is DNA sequencing, since it will undoubtedly lead to a new understanding and control of genetics and biology in a way that will alter what we mean by "life." In a sense, this invention is closely tied to the discussion above. Our ability to manipulate our own biology will play a crucial role in our survival as a species, regardless of what other developments occur.

LAWRENCE M. KRAUSS, Ambrose Swasey Professor of Physics and chairman of the Physics Department at Case Western Reserve University, is the author of *The Fifth Essence, Fear of Physics, The Physics of Star Trek*, and *Beyond Star Trek*.

GINO SEGRE

Lenses

M y choice for the greatest invention of the past two thousand years is the lens. First of all, without lenses you might not be able to read this—and, even worse, you might never have been able to read anything at all, if your vision had needed correcting. I remember Teddy Roosevelt's description of getting his first pair of glasses and suddenly having the world come into focus. Seeing clearly is, of course, no small matter, but (with apologies to Nicholas Humphrey) it seems limited to pick eyeglasses as the greatest invention of the past two thousand years, so my vote is for lenses big and small, alone and combined: the lenses we use to read the universe and the intricacies of life are variations of those we use to read the written word.

I am going to start, however, with plain old spectacles. We don't really know when they first began to be used. They were not uncommon in fourteenth-century Italy, and by 1600 there were specialized artisans who carefully ground lenses, keeping their tricks secret. One of them, a Dutch spectacle maker named Lippershey, noticed that a combination of two lenses made distant objects bigger. He tried to use this discovery to get rich. He didn't succeed, but several of his two-lens devices were made. By 1609 one of them had reached a transplanted Florentine named Galileo Galilei, who was teaching at the University of Padua. He pointed his device—or telescope, as it was later called—at the night sky and looked out. He took his telescope apart, rebuilt it, improved it, and looked some more. What he saw changed our view of the world. The sun rotated around its axis, Venus revolved around the sun, the moon had mountains and valleys, Jupiter had four moons, and the Milky Way was made up of vast numbers of stars. It was crystal clear that the old Ptolemaic vision of the universe was wrong. Copernicus and Kepler were right, the earth was not the center of the universe, and there was no going back. We were launched on our exploration of outer space.

It is a short journey from the telescope to the microscope. Not surprisingly, they were discovered at around the same time. After all, they are both just the simple piecing together of the right two lenses in correct positions. Galileo used the telescope brilliantly, but he also peered through a microscope of sorts. He saw flies the size of sheep and spots of dirt that looked like rocks, but he did not know what to make of it. In 1665 Robert Hooke published a bestseller called *Micrographia*. The book had a series of beautiful plates in it, Hooke's rendering of what he had seen with his microscope. There was a fly's eye, mold on the leaf of a rose, a picture of a louse, and so on. All very pretty, but it did not lead to anything. The microscope was a tool in search of a problem. The problem eventually did develop, and it was nothing less than understanding the origins of life. This came into focus (no pun intended) when Anton van Leeuwenhoek in 1678 made a lens good enough to get a magnifying power close to 500. At that point, a whole rich substructure was revealed. A drop of pond water turned out to be filled with little "animalcules" swimming in it. Van Leeuwenhoek had discovered bacteria. It took another two hundred years to really understand what he had seen, but then it also took three hundred years to understand that the Milky Way was just one of many galaxies.

I have been saying that the lens is the greatest invention of the past two thousand years, but an excellent lens had already been perfected over the course of millions of years by creatures so primitive they didn't even know how to make a fire. Despite this comparative ignorance, their lenses are as good as anything we can dream of making in the lab today. Of course I am talking about our own ancestors, and the lens I am describing is our eye's lens. It was developed by that diabolically clever builder we call evolution. There are many places and ways to learn just what a good job evolution did, but my favorite was offered by Richard Feynman in a physics course he taught at Caltech. Given who Feynman was, none of the course was ordinary and some of it was extraordinary, the work of a true genius. He describes how light rays enter our eye and are immediately bent and focused toward the retina by a surface lens we call the cornea. After the first focusing, the rays travel through a chamber filled with fluid and then meet the second focuser, known simply as the lens. This lens is exquisite, a thing of beauty. It is structured like an onion—with transparent layers, slightly flatter toward the edges, that gradually vary the bending of light, all designed for optimal focusing. The curvature of the lens can be adjusted by muscles located at the sides of the lens, which delicately but firmly hold it in place. With a little luck the lens forms the perfect image on the best of all screens, the retina. The retina is wired to the visual cortex in the brain and, voilà, we see the picture. I just implied that the brain and the retina are two separate things, but it may make more sense to talk of the retina as a piece of the brain, because it does a lot of the information processing before sending on its results through the optic nerve to the cortex.

My candidate for the greatest invention of the past two thousand years is still the lens, but the greatest invention of all time is the brain—which, incidentally, has managed to figure out how to use the lens it is already hooked up to and the lens it has learned how to build in its never-ending attempt to understand the universe.

GINO SEGRE is professor of physics and astronomy at the University of Pennsylvania, where he was chair of the Physics and Astronomy Department from 1987 until 1992.

GEORGE DYSON

The Universal Turing Machine

My answer is the Universal Turing Machine—because it is universal. Not only as the archetype for digital computing as we practice it today, but as a lowest common discrete-state denominator, translating between sequence in time and pattern in space. The Universal Turing Machine is entwined with the foundations of mathematics, rich in implications for the future of biology, and suggests forms of communication that we have only just begun to explore.

Intelligent life that succeeds in extending across the cosmos (and over time) will assume digital representations—at least, in some phases of the life cycle—to facilitate electromagnetic transmission, cross-platform compatibility, and long-term storage. This requires a local substrate. And we are doing our best—thanks to the proliferation of our current instantiation of the UTM known as the PC—to help. When we establish contact with such an intelligence, will we receive instructions for building a machine to upload Jodie Foster? Probably not. The transmission will proceed the other way. To paraphrase Marvin Minsky, instead of sending a picture of a cat, they will send the cat itself.

GEORGE DYSON, a historian affiliated with Western Washington University, is the author of *Baidarka* and *Darwin Among the Machines*.

———————— ✦ ————————

KARL SABBAGH

Chairs and Stairs

Clearly, none of us is playing by the rules in this game; otherwise, we would all concentrate on a few key inventions that are obviously the most important: the Indo-Arabic number system, including zero; computers; the contraceptive pill. Instead, we are all reading the suggestions so far and then trying to select something different. Because I've come in late, my friend Nicholas Humphrey has bagged my first thought—reading glasses—so I'll break the rules by choosing something that was invented more than two thousand years ago but refined over the last two thousand years.

In fact, I'll also break the rules by choosing two inventions: chairs and stairs. Apart from the fact that they rhyme, they represent an imaginative leap by realizing the value to the human anatomy of an idealized platform in space at a certain height. A platform of, say, seven inches would enable a person to ascend toward some higher objective without undue effort, but that's as far as it goes. If, from that new starting point, a further platform of the same height could be constructed, the objective could be more closely approached. The refinements have to do with the fact that the greater the height you want to reach, the larger the floor area that has to be taken up by the staircase. Landings and 180-degree turns helped to solve that problem, along with the even later invention of the spiral staircase. The consequences of stairs have obviously included greater occupation density, but they have also included the propagation of the Muslim religion by allowing muezzins to call the faithful to prayer from minarets.

As far as chairs are concerned, the same thought process was involved: recognizing the value of a platform at just above knee height and then constructing it. Portability came in at some stage as well, so that instead of finding someplace of the right height on which to perch—a wall, a rock, etc.—you could carry around with you, and position where you liked, the place to park your butt. Somehow, the height was chosen, or evolved, so that we can stay for the maximum time in a fixed position with eyes, hands, and arms free to do what eyes, hands, and arms are good at. Lying down, standing up, and squatting all get uncomfortable after a while, particularly for reading or writing—although we have to admit that medieval monks seemed to manage OK transcribing manuscripts while they were standing up.

KARL SABBAGH is a writer and television producer. His programs for the BBC and PBS have encompassed physics, medicine, psychology, philosophy, technology, and anthropology. Four of his television projects have been accompanied by books: *The Living Body; Skyscraper; 21st Century Jet: The Making and Marketing of the Boeing 777;* and (with Robert Buckman) *Magic or Medicine? An Investigation of Healing and Healers.*

◆

GORDON GOULD

Double-Entry Accounting

Here is my two cents on what is significant, in addition to all the illustrious suggestions received so far: double-entry accounting. While it is not all that sexy, it has been a significant force in shaping the West and, through the globalization of market-driven economies, the world. Invented in 1494 by a Franciscan monk named Luca Pacioli, double-entry accounting was designed to help the flourishing Venetian merchants manage their burgeoning economic empires. Today it remains the core methodology for most accounting systems worldwide. It is the DOS of money.

Based on the principle of equilibrium (the balance sheet), double-entry accounting provides both control over the internal state of an agent (in this case, an economic entity) and the necessary structures required for individual organizations to cooperate in the emergent construction of modern market economies. In other words, double-entry accounting simultaneously enables organizations to regulate themselves through internal accounting and control mechanisms while also allowing the larger economy to assess the relative health and worth of an enterprise using standardized measures. If money is the blood and markets are the circulatory system of the global economy, then double-entry accounting ledgers are the nerve cells that control and respond to changes in the flow of money.

GORDON GOULD is the president of Rising Tide Studios, the parent company of the *Silicon Alley Reporter* and the *Digital Coast Reporter*. Prior to joining RTS, he was a principal at Thinking Pictures, an interactive entertainment/database technologies company, and oversaw the Multimedia/Internet Group for Sony Worldwide Networks.

BOB RAFELSON

The Gatling Gun

Richard Gatling started with a cottonseed-sowing machine and graduated to a weapon that rotated six barrels of .58-caliber brass percussion-capped cartridges at a rate of four hundred rounds a minute. It proved its battle merit in the last days of the Civil War, during the siege of Petersburg, Virginia. In the next several decades, substantially more lethal versions were bought and used by powerful armies around the globe. The Gatling gun was the first weapon of mass destruction; as such, it has spawned the ongoing, if clumsy, debate over whether weapons ought to be banned for the sake of mankind.

BOB RAFELSON is a film director and writer-producer whose work includes *Five Easy Pieces*, *The King of Marvin Gardens*, *The Postman Always Rings Twice*, *Mountains of the Moon*, and *Blood and Wine*.

STEPHEN BUDIANSKY

The Domestication of the Horse

There is an inherent bias in all such surveys, because everyone strives to be original and surprising and so shuns the obvious but probably more correct answers—such as steel, or movable type, or antibiotics, to name but three of the more obvious inventions that have utterly transformed not only how people live but the way they experience life.

The only way I can think of being surprising is to violate John's terms and go back six thousand years. But if I will be permitted to do so, I would argue that the single invention that has changed human life more than any other is the horse, by which I mean the domestication of the horse as a mount. The horse was well on its way to extinction when it was domesticated on the steppes of Ukraine six thousand years ago, but from the moment it entered the company of human beings, the horse repopulated Europe with a swiftness that announced the arrival of a new tempo of life and cultural change. Trade over thousands of miles suddenly sprang up; communication with a rapidity never before experienced became routine; exploration of once forbidding regions became possible; and warfare achieved a violence and degree of surprise that spurred the establishment and growth of fortified permanent settlements, the seeds of the great cities of Europe and Asia. For want of the horse, civilization would have been lost.

STEPHEN BUDIANSKY, a correspondent for the *Atlantic Monthly*, is the author of *The Covenant of the Wild: Why Animals Choose Domestication; The Nature of Horses: Exploring Equine Evolution, Intelligence, and Behavior;* and *If a Lion Could Talk: Animal Intelligence and the Evolution of Consciousness.*

DAVID HAIG

The Computer

My suggestion for the most important invention of the last two millennia is the computer, because of the way it extends the capacities of the human mind for accurately performing large numbers of calculations and keeping track of and accessing vast bodies of data. Like any great invention, these enhanced abilities have a light and a dark side. As a scientist, I am now able to answer questions that could not be answered before the development of the computer. On the dark side is the loss of privacy and the enhanced potential for social control made possible by the ability to manipulate large databases of personal information.

DAVID HAIG is an evolutionary biologist and an assistant professor in the Department of Organismic and Evolutionary Biology at Harvard University.

WILLIAM H. CALVIN

Computers as Modelers of Climate

Computers—not for all the obvious reasons, but because they're the essential tool for preventing a climate-triggered collapse of civilization in the future.

Computer simulations may allow us to understand the earth's fickle climate and how it is affected by detours of the great ocean currents. These detours cause abrupt coolings—the average global temperature can drop dramatically in just a few years, with droughts that set up El Niño–like forest fires even in the tropics. While volcanic eruptions and Antarctic ice shelf collapses can also abruptly cool things, what we're talking about here is a flip-flop: a few centuries later, there's an equally abrupt rewarming. This cycle has repeated every few thousand years (though it has been twelve thousand years since the most recent one).

If such events happened more gradually, taking centuries to ramp down, we could likely cope via technofixes. The big problem is not the temperature change per se, but the less-than-a-decade speed of the transition. The suddenness of the next cooling and drying is sure to set off massive warfare over remaining resources.

While these worldwide coolings have been commonplace in the past, they are not any more inevitable than local floods are; we may be able to stabilize things if we learn enough about the nonlinear triggers. Everything we know about the geophysical mechanisms suggests that another abrupt cooling could easily happen—indeed, that our greenhouse-effect warming could trigger an abrupt cooling in several different ways.

The best understood part of the flip-flop tendencies involves what happens to the warm Gulf Stream waters off Ireland, once they split into the two major branches of the North Atlantic Current. They go on to sink to the depths of the Greenland-Norwegian Sea and the Labrador Sea. That's because the cold, dry winds from Canada evaporate so much warm water that the surface waters become cold and extra salty, tending to sink through the deeper waters. At some sinking sites, giant whirlpools more than nine miles in diameter carry surface waters down into the depths. Routinely flushing the cold waters in this manner makes room for more warm waters to flow far north—and that makes Europe much more agriculturally productive than Canada or Siberia.

But this flushing mechanism can fail if fresh water accumulates on the surface, diluting the dense waters. The increased rainfall that occurs with global warming causes more rain to fall into the oceans at the high latitudes. Ordinarily, rain falling into the ocean is not considered a problem, but at these sites in the Labrador and Greenland-Norwegian Seas it can be catastrophic. So can meltwater from the nearby Greenland ice cap, especially when it comes out in surges (which happens when ice blocks a fjord, backing up a year of meltwater, and then the ice dam collapses in a day). By shutting down the high-latitude parts of this "Nordic heat pump," such consequences of global warming can abruptly cool Europe's climate. If Europe's agriculture reverted to the productivity level of Canada's (at the same latitudes, but the winds off the Pacific Ocean are not heated in the Atlantic manner), twenty-two out of twenty-three Europeans could starve.

The surprise is that it isn't just Europe that gets hit hard. Most of the habitable parts of the world have similarly cooled during past episodes—probably via changes in water vapor, the major greenhouse gas. Another failure of the Nordic heat pump would cause a population crash that would take much of civilization with it, all within a decade.

Ways to postpone such a climatic shift are conceivable, however; cloud seeding to create rain shadows in critical ocean locations is just one possibility. Although we can't yet do much about everyday weather or about greenhouse warming, we may nonetheless be able to stabilize the climate enough to prevent an abrupt cooling. Devising a long-term scheme for stabilizing the North Atlantic's flushing mechanism has now become one of the major tasks of our civilization, essential to prevent a crash of the human population, whose wars over food would leave a balkanized world in which people hated their neighbors for good reasons.

It remains to be seen whether humans are capable of passing this intelligence test that the climate sets for us. We are likely to understand the flip-flops only via computer models, and computer simulations of possible interventions are the key to intervening safely. But if we succeed, we may be able to keep our civilization from unraveling in another episode of cool, crash, and burn.

WILLIAM H. CALVIN is a theoretical neurophysiologist on the faculty of the University of Washington School of Medicine. He has written about the brain in such books as *Conversations with Neil's Brain* (with the neurosurgeon George A. Ojemann); *The Cerebral Code; How Brains Think;* and *Lingua ex Machina: Reconciling Darwin and Chomsky with the Human Brain* (with the linguist Derek Bickerton).

V. S. RAMACHANDRAN

The Indo-Arabic Number System

My favorite invention is the place-value notation system combined with the use of a symbol 0 for zero to denote a nonexistent number; this marks the birth of modern mathematics. This system was invented in India, probably during the first millennium before Christ, but was first systematized by the Hindu mathematician and astronomer Aryabhata I at the tail end of the fifth century and then transmitted to the West via the Arabs (hence the phrase "Arabic numerals"). Before this time, even simple arithmetic was tedious and time-consuming (as when the Romans and Greeks used the cumbersome "Roman numerals"—sometimes still used in the West). And math, of course, is essential for all science. Without the early invention of zero and place value, as well as the use of a symbol to denote an unknown quantity in an equation (algebra), also from India, subsequent developments could not have occurred. There would be no calculus, no Newtonian or Galilean science, no computers, and essentially no modern world.

V. S. RAMACHANDRAN is a professor of neuroscience and psychology at the University of California, San Diego, and the director of its Center for Brain and Cognition. He is the author of *Phantoms in the Brain: Probing the Mysteries of the Human Mind* (with Sandra Blakeslee).

PETER COCHRANE

The Thermionic Valve

The invention of the thermionic valve by the American inventor Lee De Forest in 1915 was the birth of the electronic age. Without this invention, most of us would never have been born. Without electronics, this planet would not be supporting the massive numbers of people now living in the West. We would not be able to communicate, compute, manufacture, or distribute atoms on the scale we now enjoy. There would be no radio, TV, computers, Internet, modern medicine, engineering, international travel of any scale, atomic power, or almost anything else we currently take for granted. In fact, our species and our civilization would have come to a halt.

The thermionic valve is very closely followed by the transistor in 1945, invented by John Bardeen and William Shockley and serving as the foundation for the personal computer.

PETER COCHRANE is Head of Research at British Telecommunication Laboratories and the author of *Tips for Time Travelers*.

HENDRIK HERTZBERG

Printing

Doesn't it seem kind of obvious that printing—under which would be subsumed all forms of large-scale reproduction of the written word, from handmade wooden type to the computer and word-processing program I'm using to write this—was the most important invention of the past two thousand years? Printing led directly to mass literacy, democracy, the scientific revolution, cyberthis, cyberthat, and lots of other good things.

A more general observation: I notice that most of the poll responses suggest that the most important invention of the past two thousand years, whatever it was, just happens to have come along in the past hundred years. Doesn't this reflect a bad case of chronocentrism—i.e., the irrational belief that one is lucky enough to be living in history's most important era? Given that people have been inventing things all along, isn't it unlikely that so many of the most important inventions would have come along in one little century out of twenty? Wouldn't it be more logical to expect them to be spaced out randomly over all twenty? Even if you correct for population growth and for the possibility that technological (unlike artistic or literary) vision gets better over time, isn't it a little myopic to imagine that the century we just happen to be living in is twenty times more inventive than any of the others? Maybe four or five times more inventive, but even that would be a stretch.

HENDRIK HERTZBERG, a senior editor of *The New Yorker*, is the author of *One Million* and a former editor of the *New Republic*.

CHARLES SIMONYI

Public Key Cryptosystems

In the spirit of completeness and risking chronocentrism big time, I nominate Public Key Cryptosystems as an invention of the last two thousand years which will remain useful long after the printing press exists only in the (electronic) history books, next to the steam engine. PKC has three amazing properties: it ensures perfect privacy, it enables perfect authentication, and it can serve as a reliable carrier of value, as gold once did. At a single stroke, PKC transformed our vision of the asymptotic result of information and networking technology from a 1984-ish nightmare to an ultimately attractive cyberspace, where identity and privacy would not be lost, despite our (and Orwell's) commonsense intuition to the contrary.

A description of PKC would be too complicated; in any case, it is available from many sources. I would rather enumerate the reasons why it seems almost magical. They have to do with numbers and mathematics.

First, it is amazing that you, a person, can "own" a number at all. This isn't a number assigned to you by the state, such as a Social Security Number or the Personal Number so popular in European countries. Nor is it a number assigned to you by nature, such as your DNA sequence, fingerprint, or any other biometric measurement expressed as a number. Nor is this a name (which is just a number) recognized by convention or by common law, such as family names or trademarks. When you pick a very, very large number, you can be sure that it will be your number alone. It is not a number dictated by your body. It is not a number given to you by some entity whose intent you may not fathom. It is a number born from your unique mind—from your very intent to possess a number.

Next, it's amazing that you can make use of your own number in public without exposing it. Common sense tells us that if the number is kept absolutely secret from anyone or any device, then it can be of no use, and that if it is exposed somehow, then it is vulnerable. After all, writing down a number is not hard at all. Once a number is written down by some entity, you own it no longer. A number and its forgery cannot be distinguished. They are equal, identical. This commonsense argument is correct about the dangers of exposing your number in any way; however, PKC has shown that common sense is wrong, in that effective use of your number can be made without exposing the number itself.

Finally, it is amazing that these uses include the provision of privacy, authentication, and value-carrying.

CHARLES SIMONYI, Chief Architect, Microsoft Corporation, focuses on Intentional Programming, a method for the creation and maintenance of computer programs.

◆

JOHN RENNIE
Volta's Electric Battery

I'd be a traitor to my inky profession if I didn't at least echo the nominations for Johann Gutenberg's movable type. But in the spirit of the game, let me throw support behind something else: Alessandro Volta's electric battery of 1800.

Static electricity was known since at least the time of the Greeks, but study of it had largely stalled. When Pieter van Musschenbroek built and discharged the first Leyden jar in 1745, nearly killing himself in the process, he also jolted the study of electricity back to life. But it was Volta's invention of a steady source of current, inspired by the electrochemical observations of Galvani, that revolutionized technology and physics. Without it, Hans Christian Oersted could not have proved that electricity and magnetism were different faces of the same force, electromagnetism. Electro-chemistry itself offered clues to the underlying electrical nature of all matter. And of course Volta's battery was the forerunner of all the electrical devices that have transformed the world over the past two centuries.

What I find so appealing about Volta's creation is that it had im-mense practical significance but also opened to us a world of phys-ical phenomena that in themselves changed our understanding of the universe. Yet it was not a bolt-from-the-blue inspiration; it pulled together other threads of discoveries by Volta's contempo-raries. There's a lesson about greatness in there somewhere.

JOHN RENNIE is the editor-in-chief of *Scientific American*.

STUART R. HAMEROFF

Anesthesia

The most important invention in the past two thousand years is anesthesia.

Have you ever had surgery? If so, either (a) part of your body was temporarily "deadened" by "local" anesthesia, or (b) you "went to sleep" under general anesthesia. Can you imagine having surgery without anesthesia—or needing surgery, or even possibly needing surgery, without the prospect of anesthesia? And beyond the agony-sparing factor is an added feature: understanding the mechanism of anesthesia is our best path to understanding con-sciousness.

Anesthesia grew from humble beginnings. Inca shamans performing trephinations (drilling holes in patients' skulls to let out evil humors) chewed coca leaves and spat into the wound, effecting local anesthesia. The systemic effects of cocaine were studied by Sigmund Freud, but cocaine's use as a local anesthetic in surgery is credited to the American ophthalmologist Carl Koller, an associate of Freud's, who in 1884 used liquid cocaine to temporarily numb the eye. Since then, dozens of local anesthetic compounds have been developed and utilized in order to temporarily block conduction from peripheral nerves and/or the spinal cord. Local anesthetics are usually injected in liquid form, and their molecules bind specifically to sodium channel proteins in the axonal membranes of neurons near the injection site, with essentially no effects on the brain. Most general anesthetics, by contrast, are gases, and they act on the brain in a remarkable fashion. The phenomenon of consciousness is erased completely, while other brain activities continue.

General anesthesia by inhalation was developed in the 1840s, using two gases previously known to be intoxicants. The soporific effects of diethyl ether ("sweet vitriol") had been recognized since the fourteenth century, and nitrous oxide ("laughing gas") was synthesized by Joseph Priestley in 1772. In 1842 Crawford Long, a Georgia physician with apparent personal knowledge of "ether frolics," successfully administered diethyl ether to James W. Venable for removal of a neck tumor. However, Long's success was not widely recognized, and it fell to a dentist, Horace Wells, to publicly demonstrate, at the Massachusetts General Hospital in 1844, the use of inhaled nitrous oxide for tooth extraction. Although Wells apparently had used the technique previously with complete success, during the public demonstration the gas-containing bag was removed too soon, the patient cried out in pain, and Wells was denounced as a fake. Two years later, however, another dentist, William T. G. Morton, returned to Mass General and successfully used diethyl ether for a tooth extraction on William Abbott. Morton called his gas "letheon," keeping its composition secret from colleagues; he was later persuaded by the Boston physician/anatomist Oliver Wendell Holmes (father of the Supreme Court justice) to use the term *anaesthesia* (later spelled *anesthesia* in the United States) to signify absence of sensation.

Although its use became increasingly popular, general anesthesia remained an inexact art; until after World War II, there were frequent deaths due to its effects on breathing and to cardiovascular collapse. Hard lessons were learned following the attack on Pearl Harbor: anesthetic doses easily tolerated by healthy patients had tragic consequences for those who were in shock because of blood loss. Advent of the endotracheal tube—which facilitates breathing and protects the lungs from stomach contents—along with anesthetic gas machines, safer anesthetic drugs, and direct monitoring of heart, lungs, kidneys, and other organ systems, have made modern anesthesia extremely safe. However, one mystery remains: exactly how do anesthetic gases work? The answer may well illuminate the grand mystery of consciousness.

Inhaled anesthetic gas molecules travel through the lungs and blood to the brain. Barely able to be contained in blood due to poor aqueous solubility, anesthetics are highly soluble in a particular lipid-like environment akin to olive oil. It turns out that the brain is loaded with such stuff, both in lipid membranes and tiny water-free ("hydrophobic") lipid-like pockets within certain brain proteins. To make a long story short, Nicholas Franks and William Lieb, at the Imperial College in London, showed in a series of articles in the 1980s that anesthetics act primarily in these tiny hydrophobic pockets in several types of brain proteins. The anesthetic binding is extremely weak, and the pockets comprise only a fiftieth of each protein's volume, so it is unclear why such seemingly minimal interactions should have significant effects.

Franks and Lieb suggest that the mere presence of one anesthetic molecule per pocket per protein physically prevents the protein from changing into the shape necessary to do its job. However, subsequent evidence has shown that certain other gas molecules can occupy the same pockets without causing anesthesia (and in fact cause excitation or convulsions). Anesthetic molecules just "being there" can't account for anesthesia. Some natural process—critical to consciousness and perturbed by anesthetics—must be happening in the pockets. What could that process be?

Anesthetic gases dissolve in hydrophobic pockets by extremely weak quantum mechanical forces known as London dispersion forces. The weak binding accounts for easy reversibility: as the anesthetic gas flow is turned off, concentrations drop in the breathing circuit and the blood, anesthetic molecules are gently sucked out of the pockets, and the patient wakes up. Weak but influential quantum London forces also occur in the hydrophobic pockets in the absence of anesthetics and govern normal protein movement and shape. A logical conclusion is that anesthetics perturb normally occurring quantum effects in hydrophobic pockets of brain proteins.

The quantum nature of the critical effects of anesthesia may be a significant clue. Several current consciousness theories propose systemic quantum states in the brain, and as consciousness has historically been perceived as the contemporary vanguard of information processing (John Brockman's "new technology = new perception"), the advent of quantum computers will inevitably cast the mind as a quantum process. The mechanism of anesthesia suggests that such a comparison will be more than mere metaphor.

STUART R. HAMEROFF, M.D., is an associate director of the Center for Consciousness Studies at the University of Arizona, in Tucson;

a professor in the university's Anesthesiology and Psychology Departments; and a clinical anesthesiologist in its Medical Center. He started the series of Tucson Conferences on interdisciplinary approaches to consciousness and is the lead editor of the series Toward a Science of Consciousness: The Tucson Discussions and Debates. See Hameroff's website: www.u.arizona.edu/~ham.

——————◆——————

JAMES J. O'DONNELL
Late-Twentieth-Century Health Care

I f you read through this growing list, you will see that people tend to believe that the most important invention in the last two thousand years is something they happen to know a lot about. Well, I know a lot about some important inventions—like the codex book (and the consequent idea that a book can be a manual for living, which leads us to the nineteenth century and its dead ends) and the computer (which gives us a model for ignoring the manual and simply living by experiment). But I think there is something far more important going on: longevity. It's not just a matter of antibiotics or anesthesia (to mention a couple of the inventions already mentioned here), but the fundamental fact that we have learned to save and extend lives by combining the scientific method with care for human beings.

A thought experiment I like to have people play is this: Review your own life and imagine what it would have been like without late-twentieth-century health care. Would you still be alive today? An astonishingly large number of people get serious looks on their faces and admit that they wouldn't. I wouldn't, that's for sure. It's medical techniques, it's antibiotics, but it's also vitamin pills and (in some ways most wondrously cost-effective of all) soap—as in the soap doctors use to wash their hands.

JAMES J. O'DONNELL, a professor of classical studies and the vice provost for information systems and computing at the University of Pennsylvania, is the author of *Avatars of the Word: From Papyrus to Cyberspace*.

STEVEN JOHNSON
The City

G iven the amount of self-reference in the answers so far, I'm tempted to nominate this very discussion list as the greatest invention of the past two thousand years, and hopefully out-meta all the other contenders.

I think part of the problem here is the fact that inventions by nature are cumulative, and so when asked to pick out the single most important one, you're inevitably faced with a kind of infinite regress: if the automobile is the most important invention, then why not the combustible engine? And so on. In that spirit—and in the spirit of nominating things you happen to be working on professionally—I'd nominate the ultimate cumulative invention: the city. Or at least the modern city's role as an information storage and retrieval device. Before there were webs and telegraphs making information faster, there were cities bringing information physically closer together and organizing it in intelligible ways. It's not a stretch to think of the original urban guilds as file directories on the storage device of the collective mind, combining disparate skills and knowledge bases and placing them into the appropriate slots. (Manuel De Landa has a wonderful riff on this in the first section of his new book, *A Thousand Years of Nonlinear History*.)

But of course the city isn't an invention proper, at least in the conventional way that we talk about inventions. It's the sum total of multiple inventions, without each of which the modern city as we know it might not exist.

I think what this discussion makes clear is that we need a better definition of "invention"!

STEVEN JOHNSON is the author of *Interface Culture: How New Technology Transforms the Way We Create and Communicate* and the editor-in-chief of *Feed* magazine.

JEREMY CHERFAS
The Basket

Some of your jump-start friends and colleagues seem to have ignored your (arbitrary?) cutoff date—so I will, too. I think you'd have to go a long way to find a more important invention than the basket. Without something to gather into, you cannot have a gathering society of any complexity—no home and hearth, no division of labor, no humanity.

This is not an original insight. I ascribe it to Glynn Isaac, a sorely missed paleoanthropologist. For me, the basket ranks right up there with hay, the stirrup, printing, and what have you.

While we're about it, though, I'd like to take issue with Dan

Dennett's choice of the battery. Granted, it has enabled all the things he says it has (and I seriously considered nominating the Walkman—a bizarre idea, the tape recorder that doesn't record—as the invention with most impact on our lives), but at what cost? All extant batteries (though not fuel cells) are inherently polluting and wasteful. It takes something like six times more energy to make a zinc-alkaline battery than the battery can store. OK, we do have rechargeable batteries, and solar panels, but they're clunky and inefficient. I can't help but think that if a small portion of the effort that has gone into inventing "better" ordinary batteries had gone into renewables, it would have markedly reduced the throw-away mentality that plagues modern civilization. As the great ecological thinker Garrett Hardin put it, once you consider the whole earth, there is no "away" in which to throw things. If we're here to ask the same question after another couple of millennia roll around, it will be because of an invention that does something about waste or how we perceive it.

JEREMY CHERFAS is a biologist and a BBC Radio Four broadcaster. His latest book is *The Seed Savers' Handbook*.

KEITH DEVLIN

The Hindu-Arabic Number System

"What is the most important invention in the past two thousand years?" is one of those questions that has no correct answer—like "What is the best novel/symphony/movie?" But if I had to make a choice, it would be the Hindu-Arabic number system, which reached essentially its present form in the sixth century. Without it, Galileo would have been unable to begin the quantificational study of nature which we now call science, and we would not have had calculus, another major invention of the period in question.

Because of its linguistic structure, the Hindu-Arabic number system allows humans who have an innate linguistic fluency but only a very primitive number sense to use their ability with language to handle numbers of virtually any useful magnitude, with as much precision as required. Today there is scarcely any aspect of life that does not depend on our ability to handle numbers efficiently and accurately. True, we now use computers to do much of our number crunching, but without the Hindu-Arabic number system we would not have any computers.

In addition to its use in arithmetic and science, the Hindu-Arabic number system is the only genuinely universal language on earth—apart, perhaps, from the Windows operating system, which has achieved the near universal adoption of a conceptually and technologically poor product by the sheer force of market dominance. By contrast, the Hindu-Arabic number system gained worldwide acceptance because it is far better designed and much more efficient for human usage than any other number system.

KEITH DEVLIN, a mathematician and science writer, is the author of many books, including *Goodbye, Descartes: The End of Logic and the Search for a New Cosmology of the Mind; Life by the Numbers; The Language of Mathematics: Making the Invisible Visible;* and *InfoSense: Turning Information into Knowledge to Survive and Prosper in the Modern Age.* He is the dean of science at Saint Mary's College of California and a senior researcher at Stanford University's Center for the Study of Language and Information.

———————◆———————

EBERHARD ZANGGER

Nothing Worth Mentioning

The tricky part of the question is not what the most important invention is, but the qualifier "in the past two thousand years." Technological innovations alter the frontier between humans and their natural habitat. Because of the insuperable importance of the environment and its resources, humans have always sought to maximize whatever advantages they can wrest from the laws of nature. As a consequence, truly fundamental innovations date back to many thousands of years ago. The most outstanding innovation of all time was probably the domestication of animals, followed by that of plants. Life in permanent homes, villages, and cities; the wheel; the sailing ship; engineering; script; conceptual achievements such as nations, democracy, religion, music and songs; even taxes, interest, and inflation—all these date back well before the beginning of the Common Era. Several innovations suggested in this forum were actually part of the everyday routines of Bronze Age people—including (for instance) language, steel, paper, and reading glasses. The scientific method must also have existed in some form, since fourteenth-century B.C. hydraulic installations in Greece perfectly meet the parameters of the given environment. Even movable type was known by 1600 B.C., as the Discos of Phaistos from Minoan Crete shows. Finally, heliocentricity was first discovered by the astronomer Aristarchos of Samos during the third century B.C.—but the concept failed peer review, and its acceptance was thus delayed by eighteen hundred years. Since the principal factors controlling people's lives today already existed two thousand years ago, the skeptic in me would intuitively vote for: Nothing Worth Mentioning.

If we were to take a stroll through a Roman town two thousand years ago—and ancient Pompeii provides a good example of a city frozen in a moment of everyday life—we would find a city containing factories (including one for fish sauce), public baths, athletic stadiums, theaters, plastered roads, proper sidewalks, pubs, homes with under-floor heating systems, and (inevitably) brothels—facilities for people who were, for the most part, in better physical shape than we are. The Roman dominion over the Western world lasted for about a thousand years, and we might still be living in the Roman era had there not been what I consider the biggest conceptualization of the past two thousand years: Jesus Christ. Fabricated for the most part—if not entirely—by the Gospels of Mark, Matthew, and Luke, and the letters of Paul, the invention of Christianity provided a common denominator uniting the many tribes suppressed by imperial control. It also formed the fountainhead of a religion that today claims more than a third of the world's population. After all, it is only because of this invention that we commemorate Y2K. Disregarding the conceptual realization of Christianity and turning toward more mundane inventions, what else distinguishes a modern city from its Roman predecessor? I think primarily electricity. Only through the invention of electricity is it possible to operate laundry machines and subnotebook computers—two inventions I personally cherish the most—as well as many of the other items suggested in this forum.

However, I recall enjoying a particularly romantic evening in the usually overcrowded, noisy Cretan tourist resort of Elounda. Some time passed before I realized what made this evening so special: a power outage had knocked out all the fluorescent lighting and the loudspeakers. Kitchen stoves and gas lanterns still worked.

This brings me back to my original response to the question: What is the most important invention in the past two thousand years? Nothing Worth Mentioning.

EBERHARD ZANGGER found out that Atlantis equals Troy. He is a geoarchaeologist and works as chief physical scientist on many archaeological field projects in Mediterranean countries. He is the author of *The Flood from Heaven: Deciphering the Atlantis Legend* and *The Future of the Past: Archaeology in the 21st Century.*

HENRY WARWICK
Nothing

A fter reading your question and the answers to it, I have my own conclusions. Allow me a few moments to digress. Basically, I'm too disengaged and alienated from the dominant cultural paradigm of messianic technology to think that there is a "greatest invention" of the past two thousand years. Every single invention has resulted either in misery or a meaningless tangent that renders it fairly useless. I don't see the human race as progressing; I see us evolving and adapting to changing times and events. We're not really "better" from age to age; we're just different.

Would I trade the time in which I live for any other? Never. But would anyone from another time trade for mine? I think not. Take a northern European medievalist, ca. 650. Would he like painless dentistry and potatoes? Surely. Would he give up a world where God reigns supreme and the universe is in complete order down to the finest detail for a world where nothing is certain, where God has been exiled to the sidelines, and where the quantity of human slaughter, poverty, and misery is unparalleled even by medieval standards? I seriously doubt that he would surrender a life of certitude and faith so that he could surf the Web and eat processed food off plastic plates.

We are all of our time; and while we can dream of other times, we always dream *in* our own time. I find it striking that Augustus Caesar would find, say, Descartes's life thoroughly comprehensible. He might not approve of the religious convictions, and he might find the clothing rather drear, but he'd recognize the basics. And in terms of almost every single aspect of life, Descartes had more to do with the time of Augustus Caesar than with mine. Transportation, communication, food—all the aspects of daily life for me—are things that Descartes would neither understand nor comprehend. It would all be magical to him. And so, too, people several hundred years hence will seem to me.

That is why I am not interested in human life extension. I don't think it's a good idea for people to live for three hundred years. Imagine if some asshole like Jefferson Davis was middle-aged now and politically active. I don't think I could hack another two hundred and twenty years of Jesse Helms or another hundred and eighty years of Adolf Hitler. So I don't think it's a good idea to evolve longer-lived people. For all the cool people who would live longer and make great art and discoveries, there would be hundreds of evil idiots rendering life miserable for everyone, and thousands of complacent dopes sitting on their hands waiting for it all to blow over—and all of them living for hundreds of years. Society would implode.

When I see an invention, I immediately see its opposite—or worse, its tangent. Plastics is a good example. Make something that will last forever. Nice idea. I can go along with that. But we make it out of something so overexploited (oil) and hence abundant that the price drops to negligible levels, perturbed only by transient political moments. And then we use it for the most trivial purposes imaginable. The use of a nonrenewable resource for "disposable products" is something I find especially galling. Millions of years' worth of ancient algae left to molder in a landfill in New Jersey as a sandwich baggie or an underarm-deodorant dispenser. Brilliant!

Another example: Aluminum. There was a time when aluminum was worth more than gold or platinum. Why? Because it was really hard to make. But today you can go to a hot dog vendor who'll dish you up some instant processed animal fat wrapped in aluminum foil. And you'll eat the hot dog and toss the foil into the trash, and it, too, ends up in the landfill. So, is the post–Iron Age such a great idea?

The printing press = World Weekly News Video = Barney the Purple Dino. The internal combustion engine = the Stupid Useless Vehicle. Radium = Chernobyl. The computer = Finnish pornography websites. Anesthesia = face lifts for aging film stars. And so on and on. Even key concepts are of no value. One of my favorites is Pythagoras. He saw the universe as geometric, and we've been spending thousands of years trying to prove him correct and have yet to do it. None of our systems of abstraction have brought us to the truth—and by definition can't. We look at our past achievements—pyramids, arithmetic, physics, yellow plastic corn forks—and think, Yes, the world is understandable! But we have no proof of such. There has been no proof that the universe is comprehensible. We have faith that it is. We are forced to believe that our endeavors and concepts cohere with reality, that the universe is mathematical, that we can understand its workings, for any belief to the contrary—well, there lies the path to madness, usually.

As you pointed out in your book *By the Late John Brockman*, science has proved one thing: nothing. If only all the rest would follow from that conclusion. Unfortunately, reality is so complex and the mind is so limited that I've been forced to conclude from the evidence that it all will always be beyond our grasp, and that we are lucky to be left with a sense of wonder and awe.

Stephen Hawking's ideas will seem antiquated in two hundred years. People will look upon Bill Gates as some atavistic Medici in blue jeans. Everything we think and imagine will be superseded and forgotten, or simply forgotten straight away.

You may think I'm some sort of thoroughgoing nihilist, but actually I'm not. I want science to keep pushing boundaries and exposing more mystery. I want the human project to continue (albeit with a tenth as many people involved), and I believe that the course of human social evolution is the most critical aspect of our species' development and always will be. Even the simplest things we do have far-reaching effects; and while not all of them are positive, they all keep the game going somewhere. I just hope I can contribute something worthwhile—something that comes out on the positive side of the balance sheet. If I can do that, then I've done my job here. I realize that that implies a contradiction in my position, but I am capable of living with contradictions.

So, what has been the greatest invention of the past two thousand years?

There are none that are "greatest." They are all quite brilliant and utterly amazing. They are all long sharp knives at our throats, as we hang on for our dear lives to the edge of a precipice. Of course, such an elevated position affords an excellent view of everything below.

HENRY WARWICK is an artist, composer, and scientist whose formal education consists of a BFA from Rutgers University in visual systems studies, a major of his own invention. He lives in San Francisco, and his works can be appreciated at www.kether.com.

PART II

How We Think

Murray Gell-Mann

Disbelief in the Supernatural

I came up with an answer to your question right away, but I am not sure if my choice is suitable. For one thing, I am not sure if it should be regarded as an invention rather than a discovery. Also, I don't know if it is really an innovation of the last two thousand years. Most likely there were many people who came to it before the year 2 B.C.; and, if so, some of them may have overcome the understandable fear of discussing their ideas publicly. As a result, we may even have documentary evidence from more than two thousand years ago.

In any case, the most important "invention" I can think of is disbelief in the supernatural—the realization that we are part of a universe governed entirely by law and chance. (Of course, the fundamental role of chance alongside law was not fully appreciated before the discovery of quantum mechanics. In that earlier era, chance could be regarded as arising entirely from ignorance of initial conditions in a fully deterministic world.)

MURRAY GELL-MANN is a theoretical physicist; recipient of the 1969 Nobel Prize in physics; a cofounder of the Santa Fe Institute, where he is a professor and co-chairman of the science board; a director of the John D. and Catherine T. MacArthur Foundation; and author of *The Quark and the Jaguar: Adventures in the Simple and the Complex*.

———————◆———————

STEVEN ROSE
Democracy and Social Justice

I don't need a full page. The answer is clear: inventions are concepts, not just technologies, so the most important inventions are the concepts of democracy and social justice and the belief in the possibility of creating a society free from the oppressions of class, race, and gender.

STEVEN ROSE, a neurobiologist, is a professor of biology and the director of the Brain and Behavior Research Group at the Open University, Milton Keynes, England. He is the author of *Molecules and Minds, The Making of Memory,* and *Lifelines;* a coauthor (with R. C. Lewontin and Leon Kamin) of *Not in Our Genes;* and the editor of *From Brains to Consciousness? Essays on the New Science of the Mind.*

———————◆———————

JOSEPH LEDOUX
Various, Including the Idea That
All People Are Created Equal

My top runners in the category of physical inventions would be ways of harnessing energy, ways of moving around the world, and ways of communicating. And since the last two depend on the first, I'd put my money on energy control and use.

But we've got lots of psychological and social inventions as well. I'd put at the top of the list the idea that all people are equal. This is an invention we could make better use of.

JOSEPH LeDoux is Henry and Lucy Moses Professor of Science in the Center for Neural Science, New York University. He is the author of the recently published *The Emotional Brain: The Mysterious Underpinnings of Emotional Life;* a coauthor (with Michael Gazzaniga) of *The Integrated Mind;* and a co-editor (with W. Hirst) of *Mind and Brain: Dialogues in Cognitive Neuroscience.*

DON GOLDSMITH

The Realization of Our Place in the Cosmos

The most important invention is a mental construct: the realization that we on earth are an integral part of a giant cosmos, not a privileged form of existence in a special place. This invention, once the province of a few intellectuals in an obscure corner of the world, has now become widespread, though it remains a minority view among the full population.

DONALD GOLDSMITH is an astronomer and the author of over a dozen books, including *Voyage to the Milky Way; Einstein's Greatest Blunder? The Cosmological Constant and Other Fudge Factors in the Physics of the Universe; The Astronomers,* the companion volume to the PBS series of the same name; and *The Hunt for Life on Mars.* In 1995, Dr. Goldsmith was the recipient of the Annenberg Foundation Award for lifetime achievement, given by the American Astronomical Society. He has also been awarded the Dorothea Klumpke-Robert Prize for astronomy popularization by the Astronomical Society of the Pacific.

STEVEN PINKER
The Alphabet and the Lens

As a cognitive scientist obsessed with language and vision, I would have to choose the inventions that have immeasurably magnified the power of these gifts: the alphabet and the lens.

The alphabet, of course, is not an invention of the last millennium, but that era has seen two grand expansions of its dominion, the printing press and digital coding (Morse, Hollerith, ASCII, EBCDIC, etc.), the basis of telegraphy and text-based computing. I can do no better in expressing the miracle of alphabet than Galileo did in 1632:

> But surpassing all stupendous inventions, what sublimity of mind was his who dreamed of finding means to communicate his deepest thoughts to any other person, though distant by mighty intervals of place and time! Of talking with those who are in India; of speaking to those who are not yet born and will not be born for a thousand or ten thousand years; and with what facility, by the different arrangements of twenty characters upon a page!

The lens similarly extends the power of our faculty of vision in time and space. Without the lens there is no light telescope, hence no modern conception of the nature of the planets, solar system, galaxies, or universe. There would be no light microscope, and hence no modern conceptions of microorganisms, cells, chromosomes, genes, or brain. There would be no still or motion photography, and hence no perception of other places and times except through the intervening minds of artists and writers.

Without these two inventions, our experience is laid out in feet and seconds; with them, it goes from angstroms to light-years, from microseconds to the age of the universe.

STEVEN PINKER is a professor in the Department of Brain and Cognitive Sciences at MIT; the director of the McDonnell-Pew Center for Cognitive Neuroscience at MIT; and the author of *Language Learnability and Language Development; Learnability and Cognition; The Language Instinct;* and *How the Mind Works.*

PAUL W. EWALD

Evolution by Selection

My nominee is the concept of evolution by selection—which encompasses natural selection, sexual selection, and the selective processes that generate cultural evolution. It offers the best explanation for what we are, where we came from, and the nature of life in the rest of the universe. It also explains why we invent and why we believe the inventions described in this list are important. It is the invention that explains invention.

PAUL W. EWALD is an evolutionary biologist and a professor of biology at Amherst College. He is the author of *Evolution of Infectious Disease.*

◆————◆

BRIAN GREENE

The Telescope

My initial thought on how to define the importance of an invention was to imagine the effect of its absence. But having just sat through yet another viewing of *It's a Wonderful Life,* I am inspired to leave contemplation of the contingencies of history to others better suited to the task. And so I will vote for my knee-jerk response: the telescope.

The invention of the telescope and its subsequent refinement and use by Galileo marked the birth of the modern scientific method and set the stage for a dramatic reassessment of our place in the cosmos. A technological device revealed conclusively that there is so much more to the universe than is available to our unaided senses. And these revelations have in time established the unforeseen vastness of our dynamic, expanding universe, shown us that our galaxy is but one among countless others, and introduced us to a wealth of exotic astrophysical structures.

BRIAN GREENE is a professor of physics and mathematics at Columbia University and the author of *The Elegant Universe: Superstrings, Hidden Dimensions, and the Quest for the Ultimate Theory.*

———————◆———————

JOSEPH TRAUB
The Scientific Method

I nominate the scientific method as the most important invention of the last two thousand years, because it makes possible the science and technology that has transformed the world. The scientific method relies on empirical data obtained by observation or experiment.

The ancient Greeks believed that the universe was an ordered structure and exhibited regular patterns, an idea first articulated by Thales of Miletus in the sixth century B.C. Aristotle (fourth century B.C.) is credited with developing a taxonomy of the natural world based on his observations. Observation also played a central role in astronomy: Hipparchus, the greatest astronomical observer of antiquity (fl. 146–127 B.C.), based his rejection of the heliocentric view of the universe on his own observations. He is best known for his discovery of the precessional movement of the equinoxes.

However, the Greeks did not conduct controlled experiments, an essential part of the scientific method in other fields of science, such as physics. This is why I consider the invention of the scientific method as falling within the time frame of the last two thousand years.

The scientific method cannot be credited to a single person or traced to a specific moment in time. Though Roger Bacon (ca. 1220–92) was a strong and early champion of experimentation, it was Francis Bacon who, in the early seventeenth century, formulated a system of empirical observation which can be roughly summarized as consisting of four essential steps: (1) identification of a problem; (2) gathering of relevant data; (3) formulation of a hypothesis from these data; and (4) empirical testing of the hypothesis.

Isaac Newton's *Philosophiae Naturalis Principia Mathematica* (1687) is commonly thought to be the culmination of the seventeenth century's scientific revolution, a great burst of systematic, ordered, and empirical science—though preceding Newton were great successes in physiology (Vesalius in Italy and Harvey in England) and astronomy (Copernicus).

Not all scientific fields were ripe for this method in the 1600s. Robert Boyle tried to bring order and clarity to significant work of the alchemists (obscured in argot), but chemistry was not yet ready for the kind of clarification other fields yielded to. Moreover, new instruments (the microscope and the telescope), together with voyages of discovery to the New World, offered a plethora of facts and specimens that could not yet be fitted into a coherent whole.

The dissemination and criticism of such work—also an essential part of the modern scientific method—now required institutions founded especially for that purpose. Before the end of the seventeenth century, societies were established all over Europe—societies that became famous for publishing scientific findings in such clear language that experiments and discoveries could be reproduced and checked by others.

Considerable controversy has arisen about the relation between theory and experiment, as illustrated by Albert Einstein's oft-quoted aphorism that the theory determines what can be observed. Nevertheless, the power of the scientific method is the engine driving much that my colleagues and I discuss in these pages.

JOSEPH TRAUB is Edwin Howard Armstrong Professor of Computer Science at Columbia University. He started his pioneering work on what is now called information-based complexity in 1959. He continues that work and also investigates matters as diverse as the computational complexity of economic models, the limits to scientific knowledge, the value of information, the value of mathematical hypotheses, and computational mathematical finance. He is the author of nine books, including *Complexity and Information*.

STANISLAS DEHAENE

The Concept of Education

The most important human invention is not an artifact, such as the pill or the electric shaver. It's an idea—the very idea that made all these technical successes possible—and that is the concept of education.

Our brain is nothing but a collection of networks of neurons and synapses—networks that have been shaped by evolution to solve specific problems. Yet, by means of education and culture, we have found ways to recycle those networks for other uses. With the invention of writing, we recycled our visual system to read. With the invention of mathematics, we applied our innate networks for number, space, and time to all sorts of problems beyond the original relevant domain. Education is the key invention that enables all these rewirings to take place at a time in life when our brains are still optimally modifiable.

As David Premack likes to remind us, *Homo sapiens* is the only primate to have invented an active pedagogy. Without education, it would take only one generation for all the inventions that others have mentioned here to vanish from the surface of the earth.

STANISLAS DEHAENE is a cognitive neuroscientist at L'Institut National de la Santé et de la Recherche Médicale, in Orsay. He studies the cognitive neuropsychology of language and number processing in the human brain, and is the author of *The Number Sense: How the Mind Creates Mathematics*.

JOHN C. DVORAK

Computer Networks

Let's ignore discoveries (germs) and technique (scientific method) for starters, before determining the greatest invention. Moreover, I think the printing press, a device invented to rip off the Bible-buying public, should be relegated to its rightful place as number two, in favor of a newer invention: computer networks. While it is easy to romanticize the past by citing the printing press, steam engines, or eighteenth-century lug nuts, we ignore the fact that our inventiveness as a civilization is increasing, not decreasing, and newer inventions might be the most important inventions. And let's choose an invention in and of itself, and not argue about derivatives. Right now, the invention that is revolutionizing the world (more than TV, for sure) is the computer network—the Internet, in particular. And for what it's worth, arguing that none of this would be possible if human beings hadn't learned to grunt first, therefore grunting is the most important invention, is nonsense.

More interesting in this artificial discussion is the fact that most of the participants, including myself, have chosen an invention from their particular specialty. Perhaps we should ask ourselves, What is the most insidious invention of the past two thousand years? How about specialization? Look at how insidious it is in this discussion. So much so that it's frightening. Change the topic! Discussing the most insidious inventions would be more fun than talking about the importance of hay, the concept of infinity, or Gödel! Just think of the possibilities: we could nominate plastic, the stock market, the vibrating dildo, sitcoms, the literary agent, Microsoft Visual BASIC, the animated cartoon, CNN, the wristwatch, roller blades, the spinach soufflé. The possibilities are endless. Let's start over.

JOHN C. DVORAK is a columnist at *PC Magazine, PC/Computing, Computer Shopper, PC-UK, Info* (Brazil), *Boardwatch, Barrons Online,* and the host of Public Radio's *Real Computing and Silicon Spin,* on ZDTV. He is the author of numerous software guides, most of them out of print, and a cofounder of the Oswald Spengler Society.

GEOFFREY MILLER

Marketing

Marketing has become the dominant force in culture. It is commonly misunderstood as a pretentious term for advertising. But it is more than that. It is a systematic attempt to fulfill human desires by producing goods and services that people will buy. It is where the wild frontiers of human nature meet the wild powers of technology. Almost everything we buy is the result of some marketing people in some company thinking very hard about how to make us happy. They do not always get it right, but they try. Production is no longer guided by the feedback provided by last quarter's profit figures—it is guided by research into human preferences and personalities, by focus groups, questionnaires, demographics. Psychology has given way to market research as the most important investigator of human nature.

Markets are ancient, but the concept of marketing arose only in the middle of the twentieth century. In agricultural and mercantile societies there were producers, guilds, traders, bankers, and retailers, but economic consciousness was focused on making money, not on fulfilling consumer desires. With the industrial revolution, mass production led to an emphasis on cost efficiency rather than consumer satisfaction. As markets matured in the early twentieth century, firms had to compete harder for market share, but they did so through advertising and sales promotions aimed at unloading goods on resistant customers.

By the time Arthur Miller's *Death of a Salesman* reached the stage at the end of the 1940s, consumer goods companies like Procter & Gamble and General Electric were developing a more respectful, inquisitive attitude toward the consumer, establishing marketing departments to find out what people wanted from their detergents and their lightbulbs. Success spawned imitators, and most corporations now have marketing departments that coordinate product research and development, advertising, promotion, and distribution.

Everything these firms do is aimed at making profits by satisfying consumers. This was the invisible revolution of the 1960s—the most important but least understood revolution in human history, marking a power shift from institutions to individuals. It did not get the same press as the civil rights movement, the new left, feminism, or environmentalism, but it radically changed the way business works, and it is still under way. The marketing orientation has become common in companies that make things for people—clothes, cars, television sets, movies—though it remains rare in heavy industry, where the immediate consumers are other businesses, and it is poorly developed in the service sector (banking, law, government, medicine, education, and the like). Once, human enterprise asked first what we could make, and second, whether anyone would want it. Now we ask first what we want, and second, how we can invent the means to get it. The shift renders most of Marx irrelevant. What can alienation and exploitation mean, when business listens so intently to our desires?

Intellectuals don't understand marketing. It is invisible to right-wing economists, who think prices carry all the information about supply and demand that markets need. There was no role for market research in Adam Smith, Friedrich Hayek, Milton Friedman, or Gary Becker. To left-wing social scientists, journalists, and Hollywood scriptwriters, marketing means nothing more than manipulative advertising by greedy corporations. Even within business, managers—although they understand marketing at a practical level—do not recognize it as a cultural, economic, social, and psychological revolution. It is not presented that way to them in business school. Business journalists still talk as though we were moving from an industrial era based on mass production to an information era based on mass entertainment. Like fish unaware of water, we do not realize that we live in the Age of Marketing. It does not much matter whether products are material or cultural, sold in stores or electronically. What matters is that products are systematically conceived, designed, tested, produced, and distributed based on the preferences of consumers rather than the convenience of producers.

How can we understand this revolution? There are two helpful historical analogies. The American and French Revolutions brought the marketing concept to politics long before it gained a toehold in business. The production-oriented state asked what taxpayers could do for it; the marketing-oriented state asks what it can do for voters. People demanded the vote so that they could tell government what services they wanted, long before they were invited to focus groups to tell manufacturers what goods they wanted. "No taxation without representation" came long before "No profits without market research."

Even before these political revolutions, the Protestant Reformation applied the marketing insight to religion. Martin Luther and John Calvin organized churches to fill the needs of worshippers dissatisfied with a production-oriented papacy that churned out costly rituals in a dead language. The three thousand denominations of Christian faith are just what we expect from efficient market segmentation, given diverse consumers of religious services. Similar shifts occurred from production-oriented Hinayana Buddhism to market-oriented Mahayana Buddhism, and from Orthodox to Reform Judaism. The common denominator in business marketing, political democracy, and religious reform is the transfer of power from service providers to service consumers.

Is the marketing revolution a good thing? On the upside, it promises a golden age in which social institutions and markets are systematically organized to maximize human happiness. Henry Ford thought he knew what people wanted in a car: something cheap, reliable, and black. Ford sold millions of Model Ts in the 1920s with this mass marketing strategy. Then General Motors came along, segmenting the market into strata according to income, age, and tastes, attracting buyers by fulfilling their needs more precisely. Now all car companies work hard to find out what people really want from cars, and they try to build cars to fit the preferences. Market research uses all the same empirical tools as experimental psychology, but with larger research budgets, better defined questions, more representative samples of people, and more impact.

On the downside, marketing is Buddha's worst nightmare. It is the Veil of Maya made scientific, backed by billion-dollar campaigns, and perpetuating the grand illusion that desire leads to fulfillment. It promotes a narcissistic lifestyle based on pleasure, social status, romance. A product's associations become more important than its qualities. Runaway marketing, not nuclear war, may be the most common extinguisher of intelligent life in the universe.

Marketing brings more immediate problems. Like democracy, it disturbs the intellectual elite, who do not always like companies and states that provide what the masses want. Many consumers may want food full of sugar, fat, and salt; cigarettes, beer, and marijuana; motorcycles and handguns; porn videos and prostitutes; breast implants and Viagra; *Baywatch* and TV gladiators; gas-guzzling, pedestrian-squashing SUVs. If non-elite citizens all voted, they might want the death penalty, prayer in schools, book burning, ethnic cleansing. Plato recognized the political tensions between democracy based on universal suffrage and the utopian visions of well-intentioned elites. His ideal of the philosopher-king was one of the first explicit rejections of the marketing orientation as a basis for society.

Marketing, like democracy, is anti-arrogance, anti-power, and anti-idealism. It replaces paternalistic progressive visions based on the illusion of popular consent with the reality of a world shaped to fulfill ordinary human desires. It is tempting to ignore the revolution, to naïvely propose that the most significant constructs of the past two millennia have been technological inventions that expand production abilities, or scientific ideas that inform elite ideals. We tend to ignore the marketing revolution because we are terrified of a world in which those ideals are powerless to control the fruits of technology. Marketing is the most important invention of the last two millennia because it is the only revolution that has ever succeeded in bringing real power to the people—not just the power to redistribute wealth, but the power to make our means of production transform the natural world into a playground for human passions. Marketing is not just the icing on the material world. It has become the recipe, the kitchen, and the cook.

GEOFFREY MILLER is an evolutionary psychologist at the Center for Economic Learning and Social Evolution (ELSE), University College, London, who specializes in research concerning evolution of the human mind through sexual selection.

LUYEN CHOU
Philosophical Skepticism

I would have to vote for philosophical skepticism as the most important "invention" (if one thinks of invention as fabrication rather than discovery, as it is more archaically meant) of the past two thousand years. The notion that there is a "truth behind" things and a "bottom" to the matter has instilled in all of us, whether scientists, philosophers, theologians, or laypeople, a maniacal obsession with improving our explanatory capabilities. As such, skepticism can be seen as the driving force behind science and technology as well as modern conceptions of faith and the soul. Of course, one might argue that skepticism has been around for longer than two thousand years; but its characterization as a

fundamental problem to be contended with before any constructive work can be done seems to me a peculiarly modern invention,
a defining feature of our intensively self-conscious, post-Cartesian
world.

LUYEN CHOU is the president and CEO of Learn Technologies Interactive, an interactive media developer and publisher in New
York City.

PIET HUT

The Construction of Autonomous Tools

The construction of autonomous tools is my candidate for the
most important invention of the past two millennia.

Artificial complex adaptive systems, from robots to any type of
autonomous agent, will change our worldview in a qualitative
way, comparable to the way in which our view of the world was
changed forever when we started to use tools.

Tinkering with tools has shaped our view of the world and of
ourselves. For example, the invention of the pump enabled us to
understand the mechanical role of the heart. Science was born
when laboratory apparatus was used to select, among mathematical theories of the physical world, the one that corresponded most
closely to reality. But all those tools have been lifeless and soulless
things, and it is no wonder that our scientific worldview has
tended to objectify everything. Grasping the proper role of the
subject pole of experience through the invention of subject-like
tools may provide the key to a far wider worldview.

With the invention of perspective in the late Middle Ages, we shifted our collective Western experience one-sidedly to the object pole, leaving the subject pole out of the picture. We started looking at the world from behind a window, and a couple of centuries later, in science, we attempted to take a god's-eye view of the world. Now we are coming full circle, with our science and technology providing us the means of exploration of the role of the subject.

We have made only the first steps toward building artificial subjects. Just as our current artificial objects are vastly more complex than some of our earlier tools, such as the wheel or the bow and arrow, our artificial subjects will grow more complex, powerful, and interesting over the centuries. But already we can see a glimmer of what lies ahead: our first attempts to build autonomous agents have taught us new concepts. As a result, we are now beginning to explore self-organizing ecological, economic, and social systems—areas of study where thinglike metaphors hopelessly fail.

PIET HUT is a professor of astrophysics at the Institute for Advanced Study, in Princeton, New Jersey. He is involved in the project at Tokyo University to build GRAPEs (short for GRAvity PipE), the world's fastest special-purpose computers, designed for large-scale simulations in stellar dynamics.

Thomas de Zengotita

Geometry

Geometry, though it predates the time frame of your poll, still has to be my candidate for "greatest invention." *Ge* for "earth," *metron* for "measure"—the original Greek says it all. Before geometry, there was tinkering. After geometry, the world lay exposed to theory (Greek for "spectacle")—and thus to human manipulation.

But its effects were not realized practically, as opposed to conceptually, until the seventeenth century, when the "natural philosophers" launched modern science and technology explicitly in programs articulated by Descartes and Bacon. Conceived in the epistemological prison of the cogito and dedicated to the proposition that "the veil of ideas" could be rent by experiment in the service of "the new reason," they set out to rend the veil and fashion the world according to their own designs on the basis of knowledge thus acquired. For that veil—the data of subjective experience—was understood to register systematically the effects of unknown objective causes out there in the world beyond the human senses: unknown causes but not unknowable, and why not? Because among the data of subjective experience certain features stood out as privileged. Shape, motion, number, and position, for example—unlike color and smell and texture—were features of subjective experience that were measurable and that happened to be essential to the very concept of a material thing existing in space. Hence, otherwise unreachable physical things that were logically guaranteed to exist out there as causes of our systematically measurable subjective experiences were also guaranteed to be describable in terms of just those measurable dimensions, those same geometries. Molecules might be invisible, but they had shape and motion and location, just like the visible entities they made up—the entities on which one might experiment. And so it was that Galileo could explain heat as a result of the motion of particles, and Boyle could construe his laws of gases; and, in general, so it was that the entire apparatus of modern physics and chemistry, with all its technological applications, was deployed along that ancient line between the intelligible and the sensible. Now it was atoms rather than Platonic forms to which Reason was conducted, but its guide was still geometry, algebraically express-ible now and supplemented by the concept of mass—but "ge" o "metry" still.

THOMAS DE ZENGOTITA teaches philosophy and anthropology at the Dalton School and at the Draper Graduate Program at New York University.

MARNEY MORRIS

The Atomic Bomb

Well, John, you did say "most important invention," not the one we should be most proud of. The invention (and det-onation) of the atomic bomb has changed the world more pro-foundly than any other human development in the last two thousand years. In seconds, nearly two hundred thousand people were dead or dying in Hiroshima, and consciousness was forever changed on our planet. Although the arms race fueled our econ-omy for a few more decades, the bomb set into motion a "warfare stalemate." With the ability to destroy our planet within the realm of possibility, we were forced to examine our rules of war and seek new means of engagement to work out our differences. And although hundreds of wars are going on at any one time on our planet, there are checks and balances, underscored by the horror of Hiroshima and Nagasaki.

Please note that if you were to have phrased the question to include time prior to two thousand years ago, I would have suggested that our most powerful invention is song.

MARNEY MORRIS is the president of Animatrix, publishers of Sprocketworks, a next-generation interactive learning channel and CD series. She also teaches interactive design in the Engineering Department at Stanford University.

DAVID E. SHAW

The Scientific Method

I know it would probably be more helpful to add something new to this list, but I found Joe Traub's nomination (the scientific method) so compelling that I'd feel dishonest doing anything but seconding it. It's hard to imagine how different our lives would be today without the steady accrual of both knowledge and technology that has accompanied the rigorous application of the scientific method over a surprisingly small number of human generations. While the notion of formulating well-explicated, testable conjectures and subjecting them to potential refutation through controlled experimentation (and, where appropriate, statistical analysis) is now second nature to those of us who work in the sciences, it's easy to forget that we were not born with an intuitive understanding of this approach, and had we lived two thousand years ago we would never have been taught to use it. Although the apparatus of formal logic would probably rate a close second in my book, I join Joe in casting my vote for the scientific method.

DAVID E. SHAW is the chairman of D. E. Shaw & Co., Inc., a global investment bank whose activities center on the intersection between technology and finance, and of Juno Online Services, the world's second-largest Internet access provider. He also serves as a member of President Clinton's Committee of Advisors on Science and Technology, and was previously on the faculty of the Computer Science Department of Columbia University.

DAVID BERREBY

The Information Economy

My candidate would be the concept of information, as something distinct from what the information is about—information as a commodity that can be bought and sold. You could say it's an ancient invention, dating back to the day of the fleet-footed messenger, but its enormous consequences had to wait for the acceleration of information-carrying technologies like the telegraph and the Internet. We're only now witnessing the cumulative impact, as the buying and selling of information begins to outweigh the buying and selling of stuff.

This is important because people who trade in information behave more like hunter-gatherers and less like our immediate ancestors, who were chained to the plow and the factory. People who succeed in an information economy are alert and adaptable to an ever changing environment. They work in small groups. They are independent thinkers who dislike taking orders, and they are fervently egalitarian. They place their faith in face-to-face relationships, not in authority or a title. As long as humanity made its living in agriculture or industry, such traits were suppressed in favor of those more amenable to centralization, obedience to authority, long chains of command.

This epoch is coming to an end. The postindustrial West no longer values stability, steadfastness, and predictability over change, adaptability, and flexibility. You can see the egalitarian ethos in the way we are no longer awed by political power, instead seeing those who hold it as just like us. (When I was a kid, people worried about the "Imperial Presidency" becoming too grand for a democracy to support—but then, when I was a kid, an ex-wrestler could not have been elected governor of Minnesota.) That's the way foraging people behave, practicing what some anthropologists call "counterdominance" to make sure that the fastest, meanest, and strongest don't get to hog all the meat, sex, and freedom. Businesspeople often remark that their twenty-something employees can't take orders and expect to be able to dress as they please and bring their parrot to work.

All this is supposed to be a consequence of prosperity. But it seems to me that the shift is far more profound. After a seven-thousand-year detour through agriculture and industry, we are returning to the ways of our proud, individualistic, headstrong, small-group-dwelling forebears—a circumstance that will reshape the human community profoundly. And it's the move from a thing economy to an information economy that's making it happen. So I nominate the information economy as it has developed over the last five hundred years, for its tendency to nudge us (well, a lucky few of us) toward our ancestral habits of freedom and equality.

DAVID BERREBY has written about science and culture in the *New York Times Magazine*, the *New Republic*, *Slate*, the *Sciences*, and many other publications.

JOHN MCCARTHY

The Idea of Continued
Scientific and Technological Progress

The most important invention is the idea of continued scientific and technological progress. The individual who deserves the most credit for this idea is Francis Bacon. Before Bacon, progress occurred, but it was sporadic, and most people did not expect to see new inventions in their lifetime. The idea of continued scientific progress became institutionalized in the Academei dei Lincei, the Royal Society, and other scientific academies; and the idea of continued invention was institutionalized with the patent laws.

JOHN MCCARTHY, a computer scientist and one of the first-generation pioneers of artificial intelligence, is a professor of computer science at Stanford University.

DAVID G. MYERS
The Control Group

O thers in this science-minded group have appropriately mentioned the scientific method. Speaking for my discipline, let me sharpen this. When it comes to thinking smart—to sifting reality from wishful thinking—one of the great all-time inventions is the control group.

If we want to evaluate medical claims (from bloodletting to new drugs to touch therapy), or to assess social programs, or to isolate influences on human behavior, we construct a controlled reality. By random assignment, we form people into equivalent groups that either receive some experience or don't—thereby isolating the particular factor being studied. The power of the controlled experiment has meant the death of many wild and wacky claims, but also the flourishing of critical thinking and rationality.

DAVID G. MYERS is a professor of psychology at Hope College and the author of *The Pursuit of Happiness: Who Is Happy, and Why* and *The American Paradox*, and of several textbooks, including *Psychology* and *Social Psychology*.

JAY OGILVY
Secularism

O K, I'll weigh in with the invention of secularism—getting out from under the thumbs of the gods.

From all that historians and anthropologists can tell us, every ancient society worshipped some god or other. Superstition ran rampant. Human beings denied their own freedom and autonomy by praising or blaming the gods for their fates. Not until the advent of some bold minds like Ludwig Feuerbach, Karl Marx, Friedrich Nietzsche, and Sigmund Freud did it become thinkable (much less, fashionable) to preach atheism. These were inventors of a new order—one that allowed us to make up our game as we go along, unfettered by superstitions about the will of the gods or fear of their punishment.

For my part, I am appalled at how slowly this invention has been accepted. Over 60 percent of Americans still agree (somewhat, mostly, or strongly) that "the world was literally created in six days, as the Bible says"—a statistic confirmed on three successive national probability sample surveys by the Values and Lifestyles Program at SRI International, where I was director of research during the 1980s. Islam claims over a billion devotees. And I find remarkable the number of highly educated, intelligent adults who still embrace a childlike, wish-fulfilling belief in God.

Without kneeling down to positivism, or overestimating what is knowable, or underestimating the mysteries that remain lurking in the individual and social unconscious, let us nevertheless celebrate our liberation from superstition, remain humble before forces that transcend our individual egos, but accept the collective responsibilities of human freedom and sing, as my Global Business Network partner Stewart Brand did in the epigraph to the *Whole Earth Catalog*, "We are as gods, so we might as well get good at it."

JAY OGILVY is a cofounder and vice president of Global Business Network, a former director of research of the Values and

Lifestyles Program at SRI International, a former professor of philosophy at Yale University and Williams College, and the author of *Living Without a Goal* and *Many Dimensional Man*.

MILFORD H. WOLPOFF

Science

Science—because it brings us explanations of our world which we can act on—is by far the most important invention of this time. The fact that scientific explanations are usually wrong brings the partial illusion of progress, as well as tenure, which is a consequence of the various publications debating the various wrongnesses.

At its best, science works in a sort of Darwinian frame, where hypotheses are the source of variation (cleverness counts) and disproofs are the extinctions. Developments from hypotheses are the analogs of ontogeny; and there are various other processes that parallel the biological world, such as randomness (first publications carry excess influence by virtue of being first, just as Microsoft systems succeed by being the most common but not necessarily the best) and punctuated equilibrium (scientific revolutions are complete replacement events). There are even biological-like terms, like *meme*, that describe how hypotheses are transmitted. All and all, ever since well before Neanderthal times, when we hominids developed significantly complex culture and a language system to transmit it and give it existence beyond the lives of the participants, we have enjoyed (in the sense of the Chinese curse) interesting times.

MILFORD H. WOLPOFF is a paleoanthropologist and a professor of anthropology at the University of Michigan. An inventor of the multiregional hypothesis of human evolution, he is the author of *Paleoanthropology* and a coauthor (with Rachel Caspari) of *Race and Human Evolution*.

———◆———

REUBEN HERSH

The Interrogative Sentence; Space Travel

The most important invention of all time was the interrogative sentence—that is, the asking of questions.

However, your original request was for the most important invention of the last two thousand years, not of all time, so to that I would say space travel. If, as seems possible, we make our planet uninhabitable, space travel might make possible the continuation of humanity. At present this is highly speculative. Perhaps in a century or two we will know if space travel can develop sufficiently to provide a new home for *Homo sapiens*.

REUBEN HERSH is a mathematician and professor emeritus at the University of New Mexico. He is the coauthor (with Philip J. Davis and Elena Marchisotto) of *The Mathematical Experience*, which won the 1983 National Book Award.

———◆———

CHRISTOPHER WESTBURY
Probability Theory

My nomination for the most important invention of the past two thousand years is probability theory, which was mainly put together in a series of steps between 1654, when Blaise Pascal proposed a solution for splitting the pot in an unfinished game of chance, and 1843, when Antoine Cournot offered a definition of chance as the crossing of two independent streams of events.

There are several reasons why one might give probability theory the nomination. One might nominate probability theory because it provided for the first time a trustworthy tool for deciding how to apportion belief to multiple sources of evidence, allowing us to supplement the insufficient role played by logic alone in identifying and eliminating biases in our own reasoning processes. One might equally nominate it because it laid the foundation for statistical analysis, thus providing human beings with a vocabulary without which most scientific discoveries made in the last century would have been (literally) unthinkable. One might also nominate it for a third, more prosaic reason: by making it possible to offer insurance on a large scale, probability theory made feasible the extremely risky and expensive shipping expeditions that helped define the modern era.

However, I don't nominate probability theory for any of these reasons, but for a fourth, more basic reason. Probability theory had fundamental epistemological implications, whose importance is underappreciated in our time because those implications are so seamlessly integrated into the foundations of our modern worldview. Until the nineteenth century, the idea that there could exist deep regularities underlain by pure chance—regularities arising from distributions of events that were themselves the result of multifarious unmeasurable causes—was not only almost unknown but actually philosophically repugnant. Aristotle had hinted at the idea, as he seems to have hinted at almost everything. In his discussion of contingency in *On Interpretation*, he pointed out that not all potentially realizable (contingent) events need have the same probability of actually being realized. In doing so, he dealt the first blow (excepting the less useful blow dealt by appeal to pure skepticism) to the notion that the world is ultimately underlain by pure necessity. His discussion implied that the apparent world of law-obeying forms might be constructed of contingent building blocks. It required the invention of probability theory to make this radical idea generally thinkable, as it is today.

By making it possible for us to think systematically about abstract regularities, probability theory rescued humankind from three philosophical shackles that from the beginning of rational thought had held us back: the need to postulate a centralized controller that made everything come out right; the simplistic empiricist assumption that "what you see is what you get" (that is, that the proper objects of scientific study are roughly identical to the direct objects of the senses); and the idea that knowledge was an all-or-nothing affair, by definition universally and perfectly true. These philosophical shackles, though perhaps they have still not been totally removed, at least had to be loosened in order for science to get moving.

A whole new world of law-obeying objects to be studied was opened up by probability theory. Neither Darwin's theory of natural selection nor Maxwell's theory of statistical mechanics (both published in the same year, only a hundred and forty years ago) would have been thinkable before probability theory was thinkable. Without probability theory, humankind was unable even to conceive of the explanations for many (probably most) of the phenomena we have now explained.

CHRISTOPHER WESTBURY is a neuropsychologist in the Department of Psychology at the University of Alberta.

W. DANIEL HILLIS
The Clock

I agree that science is the most important human development in the last two thousand years, but it doesn't quite qualify as an invention. I therefore propose the clock as the greatest invention, since it is an instrument that enables science in both practice and temperament.

It was Galileo's observation of the constant period of the pendulum swing that paved the way for the invention of the pendulum clock by Christiaan Huygens in the seventeenth century. It is no coincidence that these events occurred at the beginning of the Enlightenment. Before the invention of the pendulum clock, the standard of accurate timekeeping was the sundial, which read variable hours that were long in the summer and short in the winter. Imagine trying to write Newton's laws of physics to a standard of time that varied with the season.

The clock, the embodiment of objectivity, paved the way for the rigor of objective science. It converted time from a personal experience into a reality independent of perception. It gave us a framework in which the laws of nature could be observed and quantified. The mechanism of the clock gave thinkers like Descartes and Leibniz a metaphor for the self-governed operation of natural law. The computer, with its mechanistic playing out of predetermined rules, is the direct descendant of the clock. Once we were able to imagine the solar system as a clockwork automaton, the generalization to other aspects of nature was almost inevitable, and the process of science began.

W. DANIEL HILLIS is a physicist and computer scientist; the vice president of research and development at the Walt Disney Company and a Disney Fellow; a cofounder and the chief scientist of Thinking Machines Corporation, where he designed and built some of the fastest computers in the world; and a co-chair of the Long Now Foundation, which is building a clock designed to last for ten thousand years. He is the author of *The Connection Machine* and *The Pattern on the Stone: The Simple Ideas That Make Computers Work.*

———————◆———————

MARY CATHERINE BATESON

Economic Man—Most Boring Invention

I'd like to comment on a celebrated observation by my father, Gregory Bateson, to the effect that humankind's most boring invention was "economic man."

We can be grateful that in this case no one has cleaned up his gender-biased language, because the concept is not and never was a gender-neutral one. The dangerous idea that lies behind "economic man" is the idea that anyone can be entirely rational or entirely self-interested. One of the corollaries, generally unspoken in economic texts, was that such clarity could not be expected of women, who were liable to be distracted by such things as emotions or concern for others. Economic man belongs with a set of pernicious and obsolete ideas separating mind from body and emotion from thought—a whole family of bad inventions! Gregory argued, with Pascal, that the heart has its reasons which the reason does not know, and that decisions are less likely to be destructive if made by whole persons. Gregory himself did not see the issue in terms of gender, but he did emphasize the way in which institutional roles lead to decisions made by part persons. In changing the way women and men have access to these roles, we are reinventing them. "Economic man" may actually become human.

MARY CATHERINE BATESON is a cultural anthropologist at George Mason University. She is the author of *Composing a Life; With a Daughter's Eye: A Memoir of Margaret Mead and Gregory Bateson;* and *Peripheral Visions: Learning Along the Way.*

JULIAN B. BARBOUR
The Bell and the Symphony Orchestra

If it had not been invented over three thousand years ago, I should have nominated the bell, but instead I choose the symphony orchestra. This is because, like the bell, it establishes a dramatic link between two seemingly disparate worlds—the material world of science and the world of the psyche and the arts. The symphony orchestra is surely important because it made possible classical music, the nomination of Howard Gardner. However, I choose it as a symbol for something that may be yet to come. What is more, I make my choice precisely because on just one point I disagree with Howard Gardner: classical music is crucially dependent on physical inventions—musical instruments. I have long been fascinated by one of the great conundrums of philosophy, a conundrum that was clearly recognized by Newton's contemporaries: If there is only a material world, characterized by the so-called primary qualities, such as extension, motion, and mass, then how are we to explain our awareness of the varieties of so many secondary qualities—colors, sounds, tastes, smells? The material world has no need of them and can never explain them. Of course, we all know that science can now demonstrate that specific sensations are correlated with physical phenomena, but a correlation is not necessarily a cause—both correlates may well have a common cause—and still less is it an explanation. How can the vibrations of catgut create in me the effect I experience when listening to Beethoven's quartets? Perhaps I am naïve, but I am also a committed scientist. I cannot be content to regard the secondary qualities as epiphenomena. I think there could be a physics, far richer than the one we know now, in which the secondary qualities are as real as, say, electric charge. The bell and the symphony orchestra call us to ponder higher things and wider possibilities, the domain where science is reconciled with the arts.

JULIAN B. BARBOUR is a theoretical physicist and the author of *Absolute or Relative Motion? Volume I: The Discovery of Dynamics* and *Time Does Not Exist: The Next Revolution in Physics.*

MARVIN MINSKY
The Identification of Smell

In his work on the foundations of chemistry, it occurred to Antoine Lavoisier (and also, I suppose, to Joseph Priestley) that the smell of a chemical was not necessarily a property of that chemical but of some related chemical in the form of a gas, which therefore could reach the nose of the observer. Thus crystals of pure sulfur have no smell, but sulfur's gaseous relatives, sulfur dioxide and hydrogen sulfide, have plenty of it. Perhaps this tiny insight was the key to the transformation of chemistry from an incoherent field into the great science of the nineteenth and twentieth centuries.

MARVIN MINSKY is a mathematician and computer scientist, Toshiba Professor of Media Arts and Sciences at the Massachusetts Institute of Technology, and a cofounder of MIT's Artificial Intelligence Laboratory. He is the author of eight books, including *The Society of Mind*.

———————◆———————

CHRISTOPHER G. LANGTON
The Telescope and the Theory of Evolution by Natural Selection

Like others who have responded, I think the choice is obvious. The remarkable thing is that "the obvious choice" is different for everyone!

To my mind, the most important inventions are those that have forced the largest changes in our worldview. On the basis of this criterion, I have to pick two: the telescope and the theory of evolution by natural selection. I pick two because it seems to me that there are two major categories of important inventions: those that increase complexity, and those that decrease complexity.

Those inventions that increase complexity open up vast new realms of data that cannot be accounted for by the existing worldview—and thereby make the universe less understandable and thus seemingly more complex. Those that decrease complexity identify a pattern or an algorithm in vast realms of data, ridding that data of a good deal of its seeming complication. These inventions force alterations to our worldview to account for previously unaccountable data, or to account for them more directly and simply, making the universe more understandable and thus seemingly less complex. The former tend to be instruments or devices—physical constructs; the latter tend to be concepts, theories, and hypotheses—mental constructs. Both qualify, to my mind, as "inventions." (To be careful, I should say that the former category of invention also involves a mental construct; a device alone is useless without the mental construct that points it in the right direction.)

In the first category, nothing rivals the telescope. No other device has initiated such a thoroughgoing reconstruction of our worldview. It has forced us to accept the earth (and ourselves) as merely a part of a larger cosmos. Of course, numerous cosmological theories besides that of the earth-centered universe existed before the invention of the telescope; but the telescope brought the flood of data that would resolve what were previously largely philosophical disputes. The microscope, a relative of the telescope, also brought us a previously unseen universe and runs a close second to the telescope on the worldview-shaking Richter scale.

In the latter category, there are many brilliant candidates; but I think Darwin's invention of the theory of evolution by natural selection outshines them all. It is perhaps the only truly general theory in biology, a field much more complex than physics. If we discover life elsewhere in the universe, evolution by natural selection is likely to be the only biological theory that will carry over from our terrestrial biology. Darwin's theory reduced tremendously the complication of zoological data. Critically, as with the telescope, it has put tremendous pressure on the earlier worldview to accommodate humankind as merely a part of a much larger nature. This pressure is still largely being resisted, but the outcome is clear.

In the decreasing-complexity category, another prime candidate is the second law of thermodynamics. Although the second law has not, perhaps, posed such a fundamental challenge to our collective worldview, it has tremendously reduced the complexity of a great body of data. And it profoundly affects the worldview of anyone who studies it in detail!

I might have nominated the computer, but I think that although it has substantially affected our daily routines, it has not yet substantially altered our fundamental beliefs. The computer is a kind of mathematical telescope, revealing to us a vast new realm of data about what kinds of dynamics follow from what sorts of rules. We are constantly discovering new galaxies of mathematical reality with computers; however, it will be a while before these empirical discoveries force a profound alteration of our worldview. I say this even though I think it quite possible that when the question of the most important invention of the last *three* thousand years is raised at the turn of the next millennium, the entity that raises the question, and the community that responds to it, may very well be descendants of today's computers.

CHRISTOPHER G. LANGTON, a computer scientist, is internationally recognized as the founder of the field of artificial life. He is the chief technology officer at the Swarm Corporation, and the editor of the journal *Artificial Life*.

CLAY SHIRKY

Gödel's Incompleteness Theorem

My vote for "The Most Important Invention in the Past Two Thousand Years" is Kurt Gödel's incompleteness theorem, or—to use its full name—"On Formally Undecidable Propositions of Principia Mathematica and Related Systems." The incompleteness theorem, published in 1931, proves that within every mathematical system you can invariably make statements that can neither be proved nor disproved within the system itself (hence "Formally Undecidable"). Put another way, no mathematical system, no matter how elegant or powerful, is complete.

This piece of mathematical jujitsu, proving unprovability, formally ended the strain of Western thought begun by Socrates and first fully fleshed out by Aristotle—the idea that intellectual inquiry will allow us to arrive at some fully knowable truth, that our picture of the universe can someday be complete. The incompleteness theorem shows us that a complete understanding of the universe is impossible, even in theory.

The ancillary effects of this revelation—a rejection of master narrative, an understanding that we will never know all the answers, an acceptance of contradiction, and an embrace of complexity—are just now making themselves felt in the dawn of the "postcomplete" world.

CLAY SHIRKY is a professor of new media in the Department of Film and Media, Hunter College.

———◆———

COLIN BLAKEMORE
The Contraceptive Pill

My choice for the most important invention? The contraceptive pill.

Why? Well, there are of course the well-rehearsed answers to that question. The pill did indeed fertilize the sexual liberation of the sixties, did stimulate feminism and the consequent erosion of conventional family structure in Western society—the most significant modification in human behavior since the invention of shamanism. It did help to change our concept of the division of labor, to foster the beginnings of an utterly different attitude toward the social role of women. But arguably the most important consequence of this relatively low-tech invention is the growing conviction that our bodies are servants of our minds, rather than vice versa.

The invention of the pill has triggered a cultural and cognitive revolution in our self-perception. It has contributed to our acceptance of organ transplantation and gene therapy, and to our present investigation of the notions of machine intelligence and (in some quarters) germ-line genetic manipulation. It has shifted our emphasis away from controlling our physical environment to controlling ourselves—our own bodies, and hence our physical destinies.

COLIN BLAKEMORE is Waynflete Professor of Physiology at Oxford, the director of the Oxford Center for Cognitive Neuroscience, and the author of *The Mind's Brain*.

◆

OLIVER MORTON
Genetic Engineering

I'm simply amazed by the lack of interest in genetic engineering. As far as I can see—and if I'm wrong, just disregard this—there are two votes for genetic sequencing, and of these Lawrence Krauss's is a sort of afterthought. There is no nomination of genetic manipulation per se, though Robert Shapiro talks about it in the context of sequencing. John Searle and Colin Blakemore also bring it up, in one case as a specific instrumentality, in the other as a parting shot.

Unlike many of the nominees, directed genetic manipulation is an invention in the truest sense: it's a body of tools and techniques conceived for an express purpose, and you can point to the people who invented it (their names are on the patents). In its basics, it is likely to be enduring. While I find it difficult (and unsatisfactory) to imagine the use of daily oral contraceptives lasting more than a century, it seems quite likely that the basic systems used to replicate and edit DNA in laboratories will stay reasonably stable, while their implementation will doubtless increase in its efficiency enormously. And genetic manipulation is likely to have an immense impact on timescales from centuries to millennia. While I can imagine futures in which no biological entities or environments, or any people, are pervasively engineered, they don't seem terribly likely, or necessarily desirable.

Obviously there are lots of ways that genetic modification could do bad things to us—but surely that's one of the basic criteria for being "important" and not a reason for shunning a technology or downplaying its significance. If we want to implement our discoveries about the biological world, directed genetic manipulation is one of the fundamental tools with which we will do it. And for those of a more conceptual bent, genetic manipulation will undoubtedly reshape all our discourse about nature and our place in it, as it is already doing.

I suppose if people apply a discount rate when assessing an invention's future importance, then something that is awesome in its possibilities, but new, may lose out compared with something old and momentous. The printing press's five centuries of influence might thus deserve to win out over genetic modification's couple of decades. But this approach seems wrong on two counts. First, it privileges this moment above others; second, if the future is discounted, then all this stuff about computers (and, indeed, oral contraceptives) has to go on the back burner. Both have had great impacts (though in many places the pill is used by only a minority of women, often a very small one, and may in some ways be less important than safe and legal abortion), but—especially in the case of the computer, its networks, its cryptographic potential and so on—it appears that the invention's true importance lies ahead. The same is just as true of the ability to redesign living beings.

I'm not saying that directed genetic manipulation is more important than democracy or the notion of equality or many of the other great things on the list. But it is more obviously what most people mean by an invention, and I really was surprised not to see it all over the place.

OLIVER MORTON is a freelance writer and a contributing editor at *Wired* and *Newsweek International*. He is the former editor of *Wired UK* and the former science and technology editor of the *Economist*.

———◆———

JOHN HENRY HOLLAND
Board Games

B oard games, more than any other invention, foretell the role of science in understanding the universe through symbolic reasoning. Their essence is a simple set of rules for generating a complex network of possibilities by manipulating tokens on a reticulate board.

Board games are found as artifacts of the earliest Egyptian dynasties, so they don't truly fall within the two-thousand-year limit, but they have undergone a rapid "adaptive radiation" in the last millennium. Thales' invention of logic (the manipulation of abstract tokens under fixed rules) was likely influenced by a knowledge of board games, and board games offered an early metaphoric guide for politics and war in both the East (Go) and the West (Chess). These insights, in turn, had much to do with the transition from the belief that the world around us is controlled by the whims and personalities of gods to the realization that the world can be described by laws. In the nineteenth and twentieth centuries, board games became the inspiration for mathematical models and simulations of everything from genetics and evolution to markets and social interaction. Board games also offer a simple example of the recondite phenomenon called *emergence*—"much coming from little"—as when a fertilized egg yields a complex organism consisting of tens of billions of cells. And via a mutation into video games, board games offer the next generation an entry into the world of long horizons and rigorous thought—both in short supply in the current generation.

JOHN HENRY HOLLAND is a professor of computer science and of psychology at the University of Michigan at Ann Arbor. The recipient of a MacArthur Foundation award, he is credited with the discovery of genetic algorithms, a problem-solving method that simulates the evolution of sexually reproducing organisms. He is a leading exponent of complexity theory at the Santa Fe Institute and the author of *Hidden Order: How Adaptation Builds Complexity* and *Emergence: From Chaos to Order.*

JARON LANIER

The Human Ego

Joe Traub has already nabbed the invention I would have chosen—empirical method. So I'll stake out a different claim. For present purposes, I'll claim that the most significant invention of the last two thousand years is the human ego.

The ego I'm talking about is the self-concerned human that Harold Bloom credits Shakespeare with having invented. It's what William Manchester finds definitively missing in the medieval mind. Jostein Gaarder, in his children's philosophy novel *Sophie's World,* blames Saint Augustine for inventing it. It's what the fuss is about in Nietzsche. It's what exists in existentialism.

In truth, I'm not entirely convinced that I don't find good evidence of this creature in pre-Christian/Common Era texts. (Thomas Cahill thinks it was a gift from the Jews.) But it does seem that the sense of individual self, outfitted with self-proclaimed moral responsibility, free will, consciousness, and (most important) neurotic self-obsession, at one time did not exist, and then did.

One could argue that the ego had to precede empirical method. The shift from pure rationality to empiricism relied on an acknowledgment of differing perspectives of observation (while pure rationality was thought to be independent of personal perspective). So an insular self was needed as a starting point from which to pose theories and make measurements in order to test them. Only an ego can have imperfect enough knowledge to make mere guesses about what's going on in the universe—and the hubris to test and improve those guesses.

Perhaps the ego will disappear, just as it once appeared. It is under constant attack by cybernetically inclined members of the *Edge* community. It is derided as an illusion, a source of confusion, a wishy-washy mystical delusion. The ego/self might be the final shore to be conquered in a drama reminiscent of Manifest Destiny. First Galileo blasted the belief that the earth was at the center of the universe, then Darwin smashed the idea that humans were uniquely positioned at the center of life, and now the Self itself is finally to be disabused of its presumption to centrality.

If the ego/self once did not exist, then it is not immutable. One naturally wonders whether the cybernetic metaphor can kill it. Perhaps in a hundred years people will not think of themselves as selves. Perhaps they will be far more aware of the competing internal processes in the brain, or of multiperson or artificial intelligence processes in the net, than of the old-fashioned singular self.

It might turn out that the ego was a natural inhabitant only of approximately the last two thousand years.

I personally hope the ego survives the computer.

JARON LANIER, a computer scientist and musician, is a pioneer of virtual reality. He is currently the lead scientist for the National Tele-Immersion Initiative, a consortium of universities studying the implications and applications of next-generation Internet technologies.

ESTHER DYSON

Self-Government

My choice: the notion that people can govern themselves rather than being governed by someone who claims divine right. That leads to a host of consequences: People start asking what rules make sense. Who gets to rule whom? How is power created and how can it be constrained? Now we have interesting situations where the powers of governments and of corporations are evenly matched . . . and meanwhile governments are also facing erosion of their power by the Internet, which allows at least some people to move their activities (financial transactions, opinion posting) around the world relatively freely.

ESTHER DYSON is the chairwoman of EDventure Holdings and the editor of *Release 1.0*. Her PC Forum conference is an annual industry event. She is the author of *Release 2.0: A Design for Living in the Digital Age*. (*Release 2.1*, the paperback upgrade, is now available.) She is also the interim chair of ICANN, the Internet Corporation for Assigned Names and Numbers; a director of the Electronic Frontier Foundation and the Santa Fe Institute; and a member of the President's Export Council Subcommittee on Encryption.

JOHN MADDOX

The Calculus

I'm amazed that fellow beneficiaries of this site are making such heavy weather of your premillennial assignment. Incidentally, surely some have bent your rules, in that assorted Sumerians, Assyrians, and Egyptians—not to mention Chinese, Greeks, and Romans—were well into the recording of history long before two thousand years ago.

Abreacting a little, I was tempted to suggest the central locking systems on modern motorcars as the greatest contribution to the convenience of modern life, but that's a trivial invention.

In any case, there's no doubt in my mind that the invention of the differential calculus by Newton and, independently, Leibniz, was the outstanding invention of the past two thousand years. The calculus made the whole of modern science what it is. Moreover, this was *not* a trivial invention. Newton knew that velocity is the rate of change (with time) of distance and that acceleration is the rate of change of velocity (with time), but it was far from self-evident that these quantities could be inferred from the geometrical shapes of Kepler's orbits of the planets. Nowadays, of course, mere schoolboys (and girls) can play Newton's game—it's just a matter of "changing the variables," as they say.

In the seventeenth century, it was far from obvious that the differential calculus would turn out to be as influential as later events have shown. Indeed, Daniel Bernoulli claimed that Newton had deliberately hidden his "method of fluxions" in obscure language so as to keep the secret to himself. But Leibniz's technique was hardly transparent; it fell to Bernoulli himself to interpret the scheme, much as Freeman Dyson made Feynman's electrodynamics intelligible in the 1940s.

Both Newton and Leibniz appreciated that the inverse of differentiation leads to a way of calculating the "area under a curve" (on which Newton had earlier spent a great deal of energy), but it was Leibniz who invented the integral sign now scattered through the mathematical literature. That these developments transformed mathematics scarcely needs assertion.

But the effect of the calculus on physics, and eventually on the rest of science, was even more profound. Where would field theories of any kind—from Maxwell and Einstein to Schrödinger/Feynman/Schwinger/Weinberg and the like—be without the calculus?

One can say much the same about the invention of arithmetic, but that long predates two thousand years ago. The calculus was the next big leap forward.

JOHN MADDOX is a physicist, editor emeritus of *Nature*, and the author of *Revolution in Biology*, *The Doomsday Syndrome*, *Beyond the Energy Crisis*, and *What Remains to Be Discovered*.

◆

BART KOSKO

The Calculus

M ost important invention: Calculus.
The world today would be very different if the Greeks

and not Newton/Leibniz had invented—or "discovered"—calculus. Our world might have occurred a millennium or two earlier.

Calculus was the real fruit of the Renaissance. It began when thinkers took a fresh look at infinity—at the infinitely small rather than the infinitely large. And it led in one stroke to two great advances: it showed how to model change (the differential equation) and it showed how to find the best solution to a well-defined problem (optimization). The first advance took math from static descriptions of the world to dynamic descriptions that allowed things to change or evolve in time. This is where "rocket science" became a science. The second advance had an even more practical payoff. It showed how to minimize cost or maximize profit. Thomas Jefferson claimed to have used the calculus this way to design a more efficient plow. Someday we may use it to at least partially design our offspring to minimize the effects of disease or (God forbid) to maximize good behavior.

Calculus lies at the heart of our modern world. Its equations led to the prediction of black holes and gravity waves. We built the first computers to run simpler calculus equations to predict where bombs would land. The recent evolution of calculus itself to the random version called stochastic calculus has allowed us to price the mysterious financial derivatives contracts that underlie the global economy. Calculus has led us from seeing the world as what Democritus called mere "atoms and void" to seeing the world as atoms that move in a void that moves.

BART KOSKO is a professor of electrical engineering at the University of Southern California. He is the author of *The Fuzzy Future*, *Fuzzy Thinking*, and the cyberthriller novel *Nanotime*. He has also written three textbooks: *Neural Networks and Fuzzy Systems*, *Neural Networks for Signal Processing*, and *Fuzzy Engineering*. His first research paper ("Equilibrium in Local Marijuana Games") used calculus to predict the optimal amount of planted marijuana or coca shrub in a region with any mix of growers, thieves, and police agents.

---◆---

VERENA HUBER-DYSON
The Infinitesimal Calculus

My first reaction to your question was the Zero, and the next was Infinity—but my answer, in a nutshell, is the Infinitesimal Calculus.

Creating a bridge between the two archetypes of Zero and Infinity, the infinitesimal calculus makes sense of them. It has become an indispensable tool in just about every branch of engineering and science, while its conceptual justification is inextricably tied up with the foundations of mathematics. It provides a language for the formulation of problems and laws as well as a method for constructing explanations, solutions, and predictions.

It has a history: its invention in the seventeenth century by Newton and by Leibniz articulated a concept that had long been vaguely anticipated and implicitly applied. The English empiricist needed a device for describing the motions of physical bodies, an endeavor in which he succeeded with his theory of "Fluxions." The German philosopher, however, intent on coming to grips with the universal nature of mathematics, was agonizing over what he called "the labyrinth of the continuum" even after his formulation of the calculus was generally accepted; how can infinitely small quantities add up to a nonvanishing sum? Only nineteenth-century mathematics was able to put the calculus on a rigorous foundation by a scrutiny of the concepts of a limit and of infinity along two complementary paths: the Cauchy-Weierstrass method of arithmetization and Dedekind's reduction to the then nascent discipline of set theory.

Today the calculus is alive and well. Having led to the resolution of old puzzles like Zeno's paradox while raising new ones like the continuum hypothesis, its scope is still broadening. Moreover, such younger disciplines as category theory, a historical descendant and conceptual ancestor of structural algebra, and mathematical logic, which arose out of a digital (0, 1) modeling of rational argumentation, are shedding new light on its foundations. To every layman's consternation, it turns out that "nonstandard analysis" vindicates Leibniz's use of "infinitely small" nonzero quantities!

VERENA HUBER-DYSON, a mathematician, has published research in group theory and taught in various mathematics departments, including those of the University of California at Berkeley and the University of Illinois at Chicago. She is now emeritus professor of the Philosophy Department at the University of Calgary, where she taught mathematical logic and the philosophy of the sciences and of mathematics. She is the author of the monograph "Gödel's Theorems: A Workbook on Formalization" (Teubner Texte zur Mathematik, Band 122, B. G. Teubner Verlagsgesellschaft, Stuttgart and Leipzig, 1991).

———————◆———————

JOHN HORGAN

Free Will

OK, I'll bite. Has anyone nominated free will yet? The concept is more than two thousand years old, but surely it deserves consideration as one of our most important inventions ever. Almost as soon as philosophers conceived of free will, they struggled to reconcile it with the materialistic, deterministic views of nature advanced by science. Epicurus insisted that there must be an element of randomness within nature that allows free will to exist. Lucretius called this randomness "the swerve." Modern free-willers find "the swerve" within chaos theory or quantum mechanics. None of these arguments is very convincing. Science has made it increasingly clear (to me, anyway) that free will is an illusion. But—even more so than God—it is a glorious, absolutely necessary illusion.

JOHN HORGAN, a science writer, is the author of *The End of Science: Facing the Limits of Knowledge in the Twilight of the Scientific Age* and *The Undiscovered Mind: How the Human Brain Defies Replication, Medication, and Explanation.* He has written articles for *Scientific American* (where he was a staff writer for ten years), the *New York Times,* the *Washington Post,* the *New Republic,* *Slate,* the *Times* (London), *Discover,* the *Sciences,* and other publications.

TOR NØRRETRANDERS

The Mirror

The most influential invention in the past two thousand years has been the mirror. It has shown to each person how she or he appears to other persons on the planet. Before the widespread production and use of mirrors, which came about in the Renaissance, humans could mirror themselves in lakes and metallic surfaces. But only with the installation of mirrors in everyday life did viewing oneself from the outside become a daily habit. This coincided with the advent of such phenomena as table manners and clothing styles and made possible the modern version of self-consciousness—that is, viewing oneself through the eyes of others rather than just from the inside or through the eyes of God.

Thus, consciousness as we know it is an effect of an advanced mental task: the acknowledgment of the person experienced "out there in the mirror" as the one simultaneously experienced from within; the realization that the person out there in the mirror is controlled by me, in here. Many malaises of modern life have arisen from the tendency to consider the mirror image of oneself as more real than the view from within. The invention of the mirror is closely related to the problem of free will and the invention of the modern human ego, as described herein by Jaron Lanier.

This new loop of the-outside-person-viewed-by-the-inside-person recently acquired an analog: the first images of earth seen from the moon. No longer just the planet we live on, earth has become a heavenly body, one among the other celestial objects.

But perhaps a more overarching invention of the past two thousand years is the very concept of inventing: the concept that a better axe is not just a better tool but the invention of a new principle for the design of an axe. The idea that one can, in a systematic and dedicated way, seek new ways of doing old things and ways of doing entirely new things is a remarkable invention in itself. The very term *invention* is a relatively modern one; the systematic search for innovation would seem to be a spin-off of our modern, competitive economy. To consider all things objects of change and all processes open to improvement, to see natural solutions as nonoptimal and to insist that there is always a better way—that is the result of the invention of inventions.

TOR NØRRETRANDERS is a science writer and communicator based in Copenhagen and the author of *The User Illusion: Cutting Consciousness Down to Size.*

SHERRY TURKLE

The Idea of the Unconscious

My candidate would be the idea of the unconscious—the notion that what we say and do and feel can spring from sources of which we are not aware, that our choices and the qualities of our relationships are deeply motivated by our histories.

In recent years, the Freudian contribution has tended to be seen as historical—that is, something we have passed beyond—but in large part this is because popular culture has absorbed the most fundamental ideas of psychodynamics as a given. These ideas animate our understanding of who we are within our families, with our friends, and in our work. They add a dimension to our understanding of what it is to be human—an understanding that will become increasingly important as we confront a world in which artificial intelligences are increasingly presented to us and our children as candidates for dialogue and relationship and we are compelled to a new level of reflection about what is special about being a person.

SHERRY TURKLE is a professor of the sociology of sciences at MIT and the author of *The Second Self: Computers and the Human Spirit; Psychoanalytic Politics: Jacques Lacan and Freud's French Revolution;* and *Life on the Screen: Identity in the Age of the Internet.*

———————✦———————

RICHARD DAWKINS

The Spectroscope

My candidate is the spectroscope.

The telescope resolves light from very far away; the spectroscope analyzes and diagnoses it. It is through spectroscopy that we know what the stars are made of. The spectroscope shows us that the universe is expanding and the galaxies receding; that time had a beginning, and when that beginning was; that other stars are like the sun in having planets where life might evolve.

In 1835, Auguste Comte, the French philosopher and founder of sociology, said of the stars: "We shall never be able to study, by any method, their chemical composition or their mineralogical structure. . . . Our positive knowledge of stars is necessarily limited to their geometric and mechanical phenomena."

Even as he wrote, the Fraunhofer lines, those exquisitely fine barcodes precisely positioned across the spectrum—the telltale fingerprints of the elements—had been discovered. The spectroscopic barcodes enable us to do a chemical analysis of a distant star when, paradoxically (because it is so much closer), we cannot do the same for the moon. The moon's light is all reflected sunlight, and its barcodes are those of the sun. The Hubble redshift, majestic signature of the expanding universe and the hot birth of time, is calibrated by the same barcodes. The rhythmic recessions and approaches of stars, which betray the presence of planets, are detected by the spectroscope as oscillating red and blue shifts. The spectroscopic discovery that other stars have planets makes it much more likely that there is life elsewhere in the universe.

For me, the spectroscope has a poetic significance. The Romantic poets saw the rainbow as a symbol of pure beauty, which could only be spoiled by scientific understanding. This notion famously prompted Keats in 1817 to toast "Newton's health, and confusion to mathematics," and in 1820 inspired his well-known lines:

> Philosophy will clip an Angel's wings,
> Conquer all mysteries by rule and line,
> Empty the haunted air, and gnomed mine—
> Unweave a rainbow. . . .

Humanity's eyes have now been widened to see that the rainbow of visible light is an infinitesimal slice of the full electromagnetic spectrum. Spectroscopy is unweaving the rainbow on a grand scale. If Keats had known what Newton's unweaving would lead to—the expansion of our human vision, inspired by the expanding universe—he surely would not have drunk that toast.

RICHARD DAWKINS is an evolutionary biologist and Charles Simonyi Professor for the Public Understanding of Science, at Oxford University. He is a fellow of New College and the author of

The Selfish Gene, The Extended Phenotype, The Blind Watchmaker, River Out of Eden, Climbing Mount Improbable, and *Unweaving the Rainbow.*

———◆———

PHILIP ANDERSON

Quantum Theory

The question is impossible to answer by naming just one invention; you could for instance say, with some justification, "the germ theory of disease," but then that goes back to the microscope—otherwise no one would ever have seen a germ—and thence to the lens, and therefore eyeglasses may be as important as germs, and so on. But I will give you my entry. To the amazement of my colleagues, who think of me as the ultimate antireductionist, I will suggest a very reductionist idea: the quantum theory, and I emphatically include quantum field theory.

The quantum theory forces a revision of our mode of thinking, which is far more profound than Newtonian mechanics or the Copernican revolution or relativity. In a sense, it absolutely forces us not to be reductionist if we are to keep our sanity, since it tells us that we are made up of anonymous identical quanta of various quantum fields, so that only the whole has any identity or integrity. Yet it also tells us that we really completely know the rules of the game which all these particles and quanta are playing, so that if we are clever enough we can understand everything about ourselves and our world. In other words, there is no "why" question about our everyday world that the quantum theory can't answer for us—Why is the sky blue? Why is glass transparent? What holds DNA together? Why does the sun shine?—and so on. It is hard to take oneself back a hundred years and realize the extent of our ignorance then. I have a feeling that the great majority of people just have not yet grasped that the inanimate world, at least, and a great deal of the living world as well, no longer is essentially mysterious.

Note that I said "understand," not "predict." The latter is, in principle, impossible, for reasons that have little to do with the famous uncertainty principle and a lot to do with exponential explosions of computations. At every stage—from those anonymous quanta of the electron field and the quark field and so on, all the way to you and me and, for that matter, the gatepost—the world we perceive arises by "emergence," the process whereby large aggregations of matter can spontaneously exhibit properties that are not meaningful for the simpler units from which they are made. One cell isn't a tiger, nor is one atom of gold yellow and shiny. The uncertainty principle isn't a strange property of the electron—it comes from the strangeness and complication of the quantum description of the apparatus. It's really impossible to follow in detail the quantum dynamics of any macroscopic bit of matter, only averages.

I would agree with whoever said "the scientific method," if I thought that that was a single thing invented at some identifiable time—but I know too much history and see too much difference between the sociologies of the various fields.

The other really profound discovery is the molecular basis of evolution, for which probably Oswald Avery deserves more credit than anyone. Evolution itself has, like the scientific method, much too complicated a history to be classed as a single invention.

PHILIP ANDERSON is a Nobel laureate physicist at Princeton University and one of the leading theorists on superconductivity. He is the author of *A Career in Theoretical Physics* and of several books on condensed matter physics, as well as a co-editor of *The Economy as a Complex Adaptive System.*

MICHAEL NESMITH
The Copernican Theory

After reading the various answers, I'm going to sneak through the door opened by Philip Anderson and nominate a discovery instead of an invention. And it is the Copernican theory. It was a counterintuitive idea, running opposite to the interpretation of the senses (not to mention the Catholic Church). I mean, one could "see" the sun going across the sky. What could be more obvious than that? A nice move! It took a lot of intellectual courage, and taught us more than simply what it stated.

MICHAEL NESMITH is an artist, writer, and businessman, and a former member of the Monkees.

PHILIP CAMPBELL
The Printing Press

Here's my shot: Perhaps the most challengingly important inventions are those that open up new moral dilemmas and thus cause people to question whether or not the invention should have been allowed (or the precursor discovery sought) in the first place. This even applies to Howard Gardner's suggestion of classical music. I would add Theodor Adorno's statement that, in contrast to the music of some composers, it's impossible to find any evil that might have been reinforced by Mozart's music. Whereas I believe that Wagner is still banned in Israel.

But my own suggestion is closer to my professional interests. As delightfully examined in Jared Diamond's *Guns, Germs, and Steel*, writing was at least one of the most important inventions of all time; but Sumerian cuneiform is too old by three thousand years for me to offer it. So the printing press is my response to the question. After all, even the World Wide Web is just a printing press with electronic and photonic elaborations. But I can't resist looking forward at an editorial fantasy (ignoring all sober estimations of the difficulties involved): a cumulative invention that will certainly have a capacity for good and evil. To quote William Gibson's *Neuromancer:* ". . . and still he dreamed of cyberspace . . . still he'd see the matrix in his sleep, bright lattices of logic unfolding across the colorless void." No keyboard, mouse, or screen, just neural connections and a many-dimensional space of (at least) information, to explore, organize, and communicate with at will— perhaps, dare I suggest, with occasional help from an editor. I fear it's too much for me to expect, but my grandchildren could love it.

PHILIP CAMPBELL (whose oldest offspring is thirteen) was the founding editor of *Physics World* and has been the editor of *Nature* since 1995.

STEWART BRAND

Christianity and Islam

The question does most of the answering: "What is the most important invention in the past two thousand years?" That lets out agriculture, writing, mathematics, and money. Too early.

"Most important" would suggest that we look for inventions near the beginning of the period, since they would have had the most time for accumulative impact. Where did that "two thousand" come from? From the approaching Year 2000, which is a Christian Era (now referred to as the "Common Era") date—2000 C.E. That's quite a clue.

The most important cultural (hence all-embracing) invention is a religion. Only two major religions have been invented in the last two millennia: Christianity and Islam. Try to imagine the last two millennia—or the immediate present, for that matter—without them.

STEWART BRAND is the founder of the *Whole Earth Catalog,* a cofounder of The Well, a cofounder of the Global Business Network, the president of the Long Now Foundation, and the author of *The Media Lab: Inventing the Future at MIT; How Buildings Learn;* and *The Clock of the Long Now.*

MIHALY CSIKSZENTMIHALYI

Various, Including the Flag

I always liked Lynn White's story about how the stirrup revolutionized warfare and made feudal society and culture possible. Or Lefebre des Noettes's argument about how the invention of the rudder made extensive sailing possible and thus enabled the European colonization of the world. But it is sobering to note that more than a thousand years passed before we realized the impact of these artifacts. So I am not at all sure that we have, at this time, a good grip on what the most important inventions of the past two millennia are.

Certainly the contraceptive pill is a good candidate, and so is the scientific method. I am also intrigued by the effects of such inventions as the flag—a symbol of belonging which millions will follow to ruin or victory independently of biological connectedness; or the Social Security card, which signifies that we are not alone and our welfare is a joint problem for the community; or the invention of civil rights, which, however abused and misused, points us toward a notion of universal human dignity that might yet eclipse in importance all the technological marvels of the past millennium.

MIHALY CSIKSZENTMIHALYI is Davidson Professor of Management at the Claremont Graduate University, in Claremont, California. He is the author of *Flow: The Psychology of Optimal Experience; The Evolving Self: A Psychology for the Third Millennium; Creativity;* and *Finding Flow.*

LEE SMOLIN

Mathematical Representation

The most important invention, I believe, was a mathematical idea: the notion of *representation*—that one system of relationships, whether mathematical or physical, can be captured faithfully by another.

The first full use of the idea of a representation was the analytic geometry of Descartes, which is based on the discovery of a precise relationship between two different kinds of mathematical objects—in this case, numbers and geometry. This correspondence made it possible to formulate general questions about geometrical figures in terms of numbers and functions; and when people had learned to answer these questions, they had invented the calculus. Now we have come to understand that it is nothing other than the existence of such relationships between systems of relations that gives mathematics its real power. Many of the most important mathematical developments of the twentieth century—such as algebraic topology, differential geometry, representation theory, and algebraic geometry—and the most profound developments in theoretical physics are based on the notion of a representation, which is the general term we use for a way to code one set of mathematical relationships in terms of another. There is even a branch of mathematics, called category theory, whose subject is the study of correspondences between different mathematical systems. According to some of its developers, mathematics is at its root *nothing but* the study of such relationships, and for many working mathematicians category theory has replaced set theory as the language in which all mathematics is expressed.

Moreover, once it was understood that one mathematical system can represent another, the door was open to speculation about whether a mathematical system could represent a physical system, or vice versa. It was Kepler who first understood that the paths of the planets in the sky might form closed orbits, when looked at from the right reference point. This discovery of a correspondence between motion and geometry was far more profound than the Ptolemaic notion that the planetary orbits were formed by the motion of circles on circles. Before Kepler, geometry may have played a role in the generation of motion, but only with Kepler do we have an attempt to represent the orbits themselves as geometrical figures. At the same time, Galileo—by slowing motion down through the use of devices like the pendulum and the inclined plane—realized that the motions of ordinary bodies could be represented by geometry. When combined with Descartes's correspondence between geometry and number, this insight made possible the spatialization of time—that is, the representation of time and motion purely in terms of geometry. Such spatialization not only made Newtonian physics possible, it is what we do every time we graph the motion of a body, or the change of some quantity, in time. It also enabled us, for the first time, to build clocks accurate enough to capture the motion of terrestrial, rather than celestial, bodies.

The next step in the discovery of correspondences between mathematical and physical systems of relationships came with devices for representing logical operations in terms of physical motions. This idea was realized early in mechanical calculators and logic engines, but came into its own with the invention of the modern computer.

But the final step in the process begun by Descartes's analytic geometry was the discovery that if a physical system could represent a mathematical system, then one physical system might represent another. Thus sequences of electrical pulses can represent sound waves, or pictures, and all of these can be represented by electromagnetic waves. Thus we have telecommunications—certainly among the most important inventions in its own right—which cannot even be conceived of without some notion of the representation of one system by another.

Telecommunications also gave rise to a question: What is it that remains the same when a signal is translated from sound waves to electrical impulses or electromagnetic waves? We have a name for the answer—*information*—but I don't think we really understand the answer's implications. For example, some people claim not only that some physical or mathematical system can be represented in terms of another, but also that there is some coding that would permit every sufficiently complicated physical or mathematical system to be represented in terms of any other. This brings us back to Descartes, who wanted to understand the relationship between the mind and the brain. Certainly the concept of information is not the whole answer, but it does give us a language in which to ask the question—a language that was not available to Descartes. Nevertheless, without his discovery of a correspondence between two systems of relations, we would not be able to talk about "information," we would not have most of mathematics, we would not have telecommunications, and we would not have the computer. Thus, the notion of a representation is not only the most important mathematical invention, but it is also the idea that allowed us to conceive of many of the other important inventions of the last few centuries.

LEE SMOLIN is a theoretical physicist at the Center for Gravitational Physics and Geometry at Pennsylvania State University and the author of *The Life of the Cosmos.*

GEORGE LAKOFF
The Idea of an Idea

As a cognitive linguist whose job is to study conceptual systems both conscious and unconscious, I was struck by the meaning of the term *invention*.

The most concrete "inventions" proposed have been gadgets, mechanical or biological: the printing press, the computer, the birth control pill. A step away from the concrete specific technical innovations are specific technical inventions of a mental character: Gödel's theorem, Arabic numbers, the nongeocentric universe, the theory of natural selection, the theory of computation. A step away from those are the general innovations of a mental character in specific domains like science and politics—e.g., the scientific method and democracy. I would like to go a step further and talk about the invention that was causally necessary for all of the above.

The most basic fully general invention of a mental character is the Idea of an Idea.

It's a bit more than two thousand—more like twenty-five hundred—years old, at least in the West. It is an "invention" in the sense that human beings actively and consciously thought it up: to my knowledge, not every indigenous culture around the world objectifies the notion of an idea, making it a thing that can be consciously constructed.

What is required for all other human inventions is the notion that one can actively, consciously construct new ideas. We take this for granted, but it is not a natural development. Three-year-old children have lots of ideas and even make up new ones. But they do not have the Idea of an Idea—the idea that they can construct anew; they do not naturally arrive at the idea that making up new ideas is something people do. The Idea of an Idea is a cultural creation that children have to learn.

It is only with the Idea of an Idea that we get conscious specific intellectual constructions like democracy, science, the number system, the computer, the birth control pill, and so on. The Idea of an Idea is the generative notion behind the very notion of an invention and is causally necessary for all specific inventions.

GEORGE LAKOFF is a professor of linguistics at the University of California at Berkeley, where he is on the faculty of the Institute of Cognitive Studies. He is the author of *Metaphors We Live By* (with Mark Johnson); *Women, Fire, and Dangerous Things: What Categories Reveal About the Mind; More than Cool Reason: A Field Guide to Poetic Metaphor* (with Mark Turner); *Moral Politics;* and *Philosophy in the Flesh* (with Mark Johnson).

ANDY CLARK
The Digital Ecosystem

My candidate for Most Important Invention is the digital ecosystem.

A digital ecosystem is a kind of universe realized in electronic media—a universe in which we observe incremental evolution and complex interaction. The classic examples come from work on artificial life—for example, Tom Ray's Tierra project, in which strings of code compete for resources like CPU time, and in which cascades of strategies for success develop, with later ones exploiting the weaknesses and loopholes of their predecessors. (See Ray's paper entitled "An Approach to the Synthesis of Life," in *Artificial Life 2.*)

But the idea is much broader. The World Wide Web and various browser technologies have combined to create a massive digital ecosystem populated by ideas and product descriptions, whose true impact on our lifestyle is only just beginning to be felt. The human mind was never contained in the head; it has always been a construct involving head, artifacts (such as pen and paper), and webs of communication and interaction. As I argue in *Being There,* the great achievement of the human race has been to make our world smart so that brains like ours can be dumb in peace. Continuing in this vein, the development of Web and Internet technologies may well signal the next great leap in the evolution of human thought and reason. For we now have a medium in which ideas can travel, mutate, recombine, and propagate with unprecedented ease and (increasingly) across the old barriers of culture, language, geography, and central political authority.

Moreover, and in a kind of golden loop, we can now begin to experiment with the more restricted kinds of digital ecosystems so as to improve our grip on the properties of the kind of large, distributed, self-organizing (currently Web-based) digital ecosystems of which we are now a proper part. Understanding these properties is important both for policy-making (What kind of regulation creates and maintains the optimal conditions for productive self-organization in a complex and highly uncertain world?) and for moral and economic reasons. Human brains are simply bad at seeing the patterns that will result from complex, multiple, ongoing, bidirectional interactions: see, for example, the computer simulations by Mitchell Resnick (in his 1994 book *Turtles, Termites, and Traffic Jams*), of the MIT Media Lab, that show, to most people's surprise, that if each person in a group insists on having a minimum of 30 percent of his or her neighbors "the same x" as he or she is (where x can be gender, race, sexual inclination, or whatever you like), then what evolves over a short period of time is a highly segregated ecology containing a great many "all-x" neighborhoods. Perhaps if our children are able to play with very large-scale digital ecosystems—in games such as Sim City, or using new simulation-based educational resources such as Resnick's Starlogo—they may yet learn something of how to predict, understand, and sometimes avoid, such emergent patterns.

Digital ecosystems not only radically transform the space in which human brains think and reason, but also provide opportunities to help us learn to reason better about the kind of complex system of which we are now a part. The double whammy gets my vote.

ANDY CLARK is a professor of philosophy and the director of the Philosophy/Neuroscience/Psychology Program at Washington University in Saint Louis. He is the author of *Microcognition; Associative Engines;* and *Being There: Putting Brain, Body, and World Together Again.*

GEORGE JOHNSON
Mathematical Representation

Surely one of the most powerful earthly inventions has been the ability to represent any phenomenon with numbers, either analog or digital, and then use this representation to predict outcomes in the real world. This information revolution actually began before the year zero, with the Pythagoreans, and has advanced through stages that include the invention of calculus and, most recently, Boolean algebra and all the advantages of digital modeling.

And just as important has been the recent humbling realization that there are limits to this scientific cartography—that, tempting as it is, the map can never be mistaken for the real thing.

GEORGE JOHNSON is a writer for the *New York Times,* working on contract from Santa Fe, New Mexico. His books include *Fire in the Mind: Science, Faith, and the Search for Order; In the Palaces of Memory: How We Build the Worlds Inside Our Heads; Machinery of the Mind: Inside the New Science of Artificial Intelligence;* and *Strange Beauty: Murray Gell-Mann and the Revolution in 20th-Century Physics.*

HOWARD RHEINGOLD
The Evolution of Technology

The kind of thinking that makes it possible for all these people to expound upon "the most important invention in the last two thousand years" is the most important invention in the last two thousand years. There is no such thing as the single most important invention in the last two thousand years. The evolution of technology doesn't work like that. It's a web of ideas, not a zero-sum game.

Knowing how to turn knowledge into power is the most powerful form of knowledge. The mindsets, mind tools, and institutions that make technological progress possible are all part of an invisible cultural system: the system is learned, not inherent; it was invented, not evolved; it hypnotizes you to see the world in a certain way.

What we know as "technology"—the visible stuff that hums and glows—is only the physical manifestation of a specific kind of social system. That invisible system—what Jacques Ellul called "la technique" and Lewis Mumford called "technics"—emerged over the past three centuries and is more important than all the inventions it engendered.

Do we lack one important invention at a crucial time when our inventions are becoming our only evolutionary competitor? We haven't formulated and agreed upon a way of making good decisions about the powerful technologies we're so good at creating. We have a lot of the knowledge that turns knowledge into power. We need more of the wisdom that knows what we ought to do with the power of invention.

HOWARD RHEINGOLD, the founder of the webzine and virtual community *Electric Minds* and the founding executive editor of *HotWired*, the first commercial webzine, is the author of *Tools for Thought*, *Virtual Reality*, and *Virtual Communities*.

AFTERWORD
by Jared Diamond

My eleven-year-old twin sons just told me what they learned today in school. "Daddy, Johannes Gutenberg invented the printing press, and that was one of the greatest inventions in history." I learned that too when I was a child. I suspect that the same view is taught in most other American and European schools, though not in Chinese schools.

What we are told about printing is partly true, or at least defensible. Personally, I would rate it as the best single "invention" of this millennium. (But I'll explain below why "invention" isn't quite the right word for it.) Just think of its enormous consequences for modern societies. Without printing, millions of people wouldn't have read quickly, with no transmission errors, Martin Luther's Ninety-five Theses, the Declaration of Independence, the Communist Manifesto, or other electrifying ideas that produced social upheavals. Without printing, we couldn't have developed true mass democracy. Without printing, we wouldn't have modern science, which depends on far more people being in a position to assimilate past knowledge and contribute further knowledge than in Aristotle's day. Without printing, Europeans might not have spread over the globe since A.D. 1492, because consolidation of initial European conquests (like Pizarro's capture of the Inca emperor Atahualpa) required the emigration of thousands more would-be conquistadors motivated by written accounts of Pizarro's deed.

So, I agree with half of what my kids were taught—the part about the importance of printing. No other single "invention" has had anything like the same impact. But things get more complicated when you credit this invention specifically to Gutenberg, or even when you credit him with just the printing press itself. Gutenberg did much more than invent the printing press, and he did much less than invent printing. More accurate would be the following legalistic sentence: "Gutenberg played a major practical and symbolic role in independently reinventing, in a greatly improved form and within a more receptive society, a printing technique previously developed in Minoan Crete around 1700 B.C."

Why did Gutenbergian printing take off, while Minoan printing didn't? Therein lies a fascinating story that punctures our usual image of the lonely inventor-hero—Gutenberg, James Watt, Thomas Edison, and their like. Through his contribution to the millennium's best invention, Gutenberg gave us the millennium's best window into how inventions actually unfold.

Even American and European schoolkids reared on Gutenberg hagiography soon learn that China had printing long before Gutenberg, so of course he didn't invent it. Chinese printing is known to go back to around the second century A.D., when Buddhist texts on marble pillars were smeared with ink and transferred to the new Chinese invention of paper. By A.D. 868, China was printing books. But most Chinese printers carved or otherwise wrote out a text on a wooden block, instead of assembling it letter by letter as Gutenberg did (and as almost all other subsequent printers using alphabetic scripts have also done). Thus, the credit for what Gutenberg invented is also corrected from "printing" to "printing with movable type"—that is, printing with individual letters that can be composed into texts, printed, disassembled, and reused.

Have we now got the story right? No. Gutenberg doesn't deserve credit even for that. Already, around A.D. 1041, the Chinese alchemist Pi Cheng had devised movable type made of a baked clay-and-glue amalgam. Among the many subsequent inventors who improved on Pi Cheng's idea were Korea's King Htai Tjong (cast bronze type around A.D. 1403) and the Dutch printer Laurens Janszoon (wooden type with hand-carved letters around A.D. 1430). From among all those inventors it's convenient—and, I think, appropriate—to single out Gutenberg for special credit, because of his key advances that made printing much more efficient. Those include his most novel advance, the use of a press, plus several incremental improvements on existing methods: a technique for mass-producing durable metal letters, a new metal alloy for the type, and an oil-based printing ink. We also find it convenient to focus on Gutenberg for symbolic reasons: his beautiful Bible of 1455, the first completed European book still in existence, can be considered to have launched book production in the West.

But by far the earliest claimant to have invented printing with movable type is neither Gutenberg nor Pi Cheng but an unnamed printer of ancient Crete in the Minoan age. Minoan printing is attested only by a single baked-clay disk, six inches in diameter, that poses one of the most challenging mysteries in all of archaeology. Found buried deep in the ruins of a 1700 B.C. palace at Phaistos on Crete, the disk is covered on both sides with beautiful spiraling arrays of 241 symbols constituting 45 different "letters" (actually, syllabic signs), which were undeciphered until a couple of years ago and identified as an ancient form of Greek, preceding even that of Homer. Astonishingly, the symbols of the Phaistos disk weren't scratched into the clay by hand, as was true of all other ancient writing on clay, but were instead printed by a set of punches, one for each of the 45 signs. Evidently, some ancient Cretan predecessor of Pi Cheng beat him to the idea by 2,741 years. Why did Minoan printing die out, instead of spurring further developments in printing technique as Chinese and Renaissance European printing did? Why was Renaissance Europe ready to make use of the millennium's best invention, while Minoan Crete was not?

Several problems greatly restricted the value of Minoan printing. Technologically, its hand-held, hand-made punches were clumsy. (Clumsy technology also prevented the earliest internal combustion engine, invented in 1866, from being utilized to build motorcycles and cars: the engine was seven feet tall.) To make Minoan printing efficient would have required many technological advances that were not accomplished until much later, such as paper and improved inks, metals, and presses. The early Minoan writing system itself, a syllabary rather than an alphabet, was so ambiguous that it could be used for just a few kinds of texts, perhaps only tax lists and royal propaganda. As a result, hardly anybody could read Minoan writing—possibly no more than a few dozen scribes in a few palaces. Chinese printing's usefulness was similarly limited by China's own nonalphabetic writing system. The need for thousands of different pieces of movable type, bearing the thousands of separate Chinese writing signs, condemned Pi Cheng's invention to much lower efficiency than the block printing already prevalent in China.

In these respects, Gutenberg's improvements to printing in Renaissance Europe enjoyed one big advantage over Chinese printers and many advantages over Minoan printers. Unlike Chinese writing, European writing was alphabetic, making printing with only a few dozen varieties of movable type, instead of thousands, feasible. Unlike the printers of ancient Crete, Renaissance European printers could use paper (unknown in 1700 B.C.) and also vastly improved metals, inks, and presses.

I mentioned that Gutenberg provided not only the millennium's best invention but also its best window into invention's unfolding. One conclusion shining through the window is that our usual view of inventions as keys to progress often misses the point. Coming up with an invention itself may be the easy part; the real obstacle to progress may be that society lacks the capacity to utilize the invention. Printing with movable type could not have been *so* difficult to conceive, because Pi Cheng, Laurens Janszoon, and some unnamed Minoan all came up with the same idea independently. Alas, that Minoan Gutenberg was stymied by Minoan Crete's low demand for printing. Other famously premature inventions include wheels in pre-Columbian Mexico (relegated to playtoys because Mexican Indians had no draft animals) and Cro-Magnon pottery from 25,000 B.C. (What nomadic hunter/gatherer really wants to carry pots from camp to camp?) A few decades from now, magazines may run articles wondering at the inability of late-twentieth-century America to find uses for the civilian supersonic transports (SSTs) that we were perfectly capable of building.

As the history of printing teaches us, the technological breakthroughs leading to great inventions usually come from totally unrelated areas. For instance, if a queen of ancient Crete had launched a Minoan Manhattan Project to achieve mass literacy through improved printing, she would never have thought to emphasize research into cheese-, wine-, and olive presses—but those presses furnished prototypes for Gutenberg's most original contribution to printing technology. Similarly, 1930s American military planners trying to build powerful bombs would have laughed at suggestions that they fund research into anything so arcane as transuranium elements.

We picture inventors as heroes with the genius to recognize and solve their society's perceived problems. In reality, the greatest inventors have been tinkerers who loved tinkering for its own sake, and who then had to figure out what (if anything) their devices might be good for. When Otto Hahn and Fritz Strassman discovered nuclear fission in 1938, they weren't German bomb builders: they were curious German chemists tinkering with uranium. As for Gutenberg himself, we have no idea what originally motivated him, but we do know that he was a skilled metalworker associated with a goldsmiths' guild, and he was clearly a genius at playing with metals.

The prime example of tinker-driven inventing is Thomas Edison's phonograph, widely considered to be the most brilliant invention of America's most brilliant inventor. When Edison built his first phonograph in 1877, it was not in response to a growing national clamor to hear Beethoven's symphonies at home. Instead, Edison was intrigued by the challenge of building something that could capture sound. Having built it, he wasn't sure what to do with it, so he drew up a list of ten possible uses. High on his list were recording the last words of dying people, announcing clock time, and teaching spelling. When entrepreneurs instead incorporated his invention into a machine to play music, Edison objected to this debasement of his idea.

Related to our widespread misunderstanding of inventors as setting out to solve society's problems, we say that necessity is the mother of invention. Actually, invention is the mother of necessity, by creating needs that we never felt before. (Be honest: did you really feel a need for your Walkman CD player long before it existed?) Far from welcoming solutions to our supposed needs, society's entrenched interests commonly resist inventions. In

Gutenberg's time, no one was pleading for a new way to churn out book copies: there were hordes of copyists, whose desire not to be put out of business led to local bans on printing. For many decades after the first internal combustion engine was built in 1866, motor vehicles continued to languish unneeded, because the public was happy with horses and railroads, neither of which were in short supply. Transistors were invented in the United States, but the American electronics industry ignored them to protect its big investment in vacuum tube products; it was left to Sony in bombed-out postwar Japan to adapt transistors to consumer electronics products. Similar examples abound today: the Japanese continue to prefer their hideously complex writing system over the alphabet, car owners here in Los Angeles continue to oppose subways for my city, and manufacturers of typewriter and computer keyboards continue to prefer our inefficient QWERTY keyboard layout over a rationally designed layout.

All these misunderstandings about inventions pervade our national policies of science and technology. Every year, many elected officials decry some areas of basic research as a waste of tax dollars and urge that funding instead concentrate on "solving problems"—that is, on applied research. The most pernicious exponent of this philosophy was William Proxmire, the former U.S. senator from Wisconsin, who regularly announced his "Golden Fleece" awards, often singling out for ridicule some basic research project that he considered especially useless. Of course, much applied research is necessary in order to translate basic discoveries into workable products—a prime example being the Manhattan Project, which spent three years and two billion dollars to turn Hahn and Strassman's discovery of nuclear fission into an atomic bomb. But the ignorant Proxmires of the world fail to realize that neither the solutions to most difficult problems of technology nor the potential uses of most basic research discoveries have been predictable. Instead, penicillin, recombinant DNA, and all the other wonders of twentieth-century science and technology were discovered accidentally—by tinkerers driven by curiosity.

So, forget those stories about genius inventors who perceived a need of society, solved it single-handedly, and thereby transformed the world. There has never been such a genius; there have only been long processions of replaceable creative minds who made serendipitous or incremental contributions. If Gutenberg hadn't devised the better alloys and inks used in early printing, some other contemporary tinkerer with metals and oils would have done so. For the best invention of the millennium, do give Gutenberg some of the credit—but not too much.

JARED DIAMOND, a professor of physiology at UCLA Medical School, is the author of *Guns, Germs, and Steel: The Fates of Human Societies*, which won the Pulitzer Prize for general nonfiction in 1998 as well as Britain's 1998 Rhone-Poulenc Science Book Prize. He is also the author of *The Third Chimpanzee*, which won the *Los Angeles Times* Book Award for the best science book of 1992 and the 1992 Rhone-Poulenc Science Book Prize; and *Why Is Sex Fun?*

CONTRIBUTORS

Bob Rafelson (the Gatling gun), 83
V. S. Ramachandran (the Indo-Arabic number system), 89
John Rennie (Volta's electric battery), 93
Howard Rheingold (the evolution of technology), 179
Steven Rose (democracy and social justice), 112
Douglas Rushkoff (the eraser), 40
Karl Sabbagh (chairs and stairs), 80
Roger C. Schank (the Internet), 51
John R. Searle (the green revolution), 33
Gino Segre (lenses), 76
Terrence J. Sejnowski (the digital bit), 69
Robert Shapiro (genetic sequencing), 48
David E. Shaw (the scientific method), 131
Clay Shirky (Gödel's incompleteness theorem), 148
Charles Simonyi (Public Key Cryptosystems), 91
Lee Smolin (mathematical representation), 172
Dan Sperber (the computer and the atomic bomb), 59
Tom Standage (telecommunications technology), 23
Duncan Steel (the thirty-three-year English Protestant
    calendar), 62
Peter Tallack (the stirrup and the horse collar), 66
Joseph Traub (the scientific method), 117
Arnold Trehub (Otto von Guericke's static electricity
    machine), 27
Colin Tudge (the plow), 25
Sherry Turkle (the idea of the unconscious), 163
Henry Warwick (nothing), 105
Christopher Westbury (probability theory), 139
Milford H. Wolpoff (science), 137
Eberhard Zangger (Nothing Worth Mentioning), 104
Carl Zimmer (waterworks), 47